商业建筑

COMMERCE

U0824019

最新室内细部设计实例集VI

Interior Detail

商业建筑

COMMERCE

[韩] 建筑世界株式会社 编

田 华 白昭薰
张小卿 郝赤彪 译

中国建筑工业出版社

目　　录

饭店餐馆

自助餐厅与酒吧

娱乐场所

contents

Restaurant

Cafe & Bar

Recreation

铉京饭店
康乐健体韩式食堂—健
羌羌水来烤肉店

桃梨园
辣不理餐厅

陶茂饭店
阿巴洛克餐馆

最新室内细部设计实例集Ⅵ

饭店餐馆

Restaurant

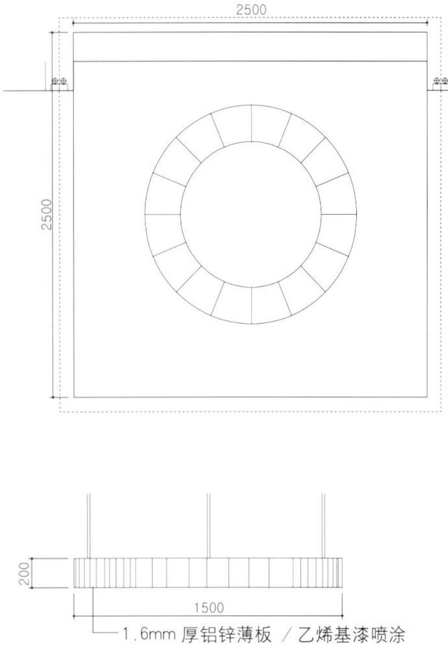

灯箱盒一平面和立面图

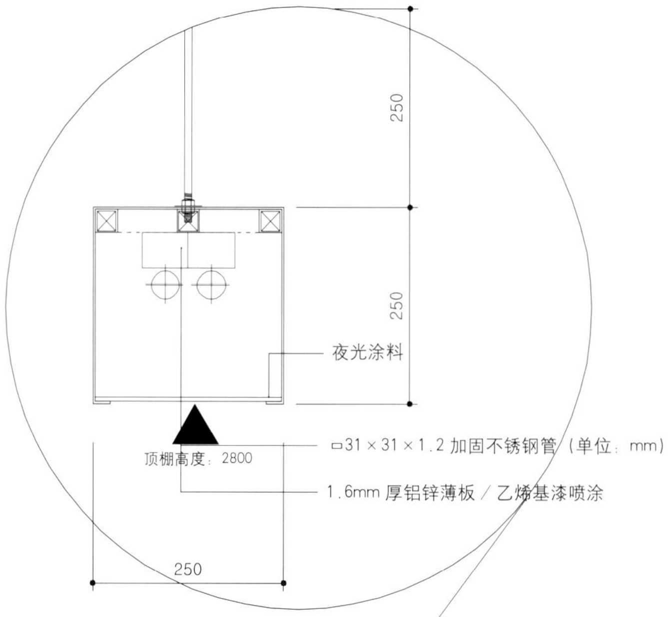

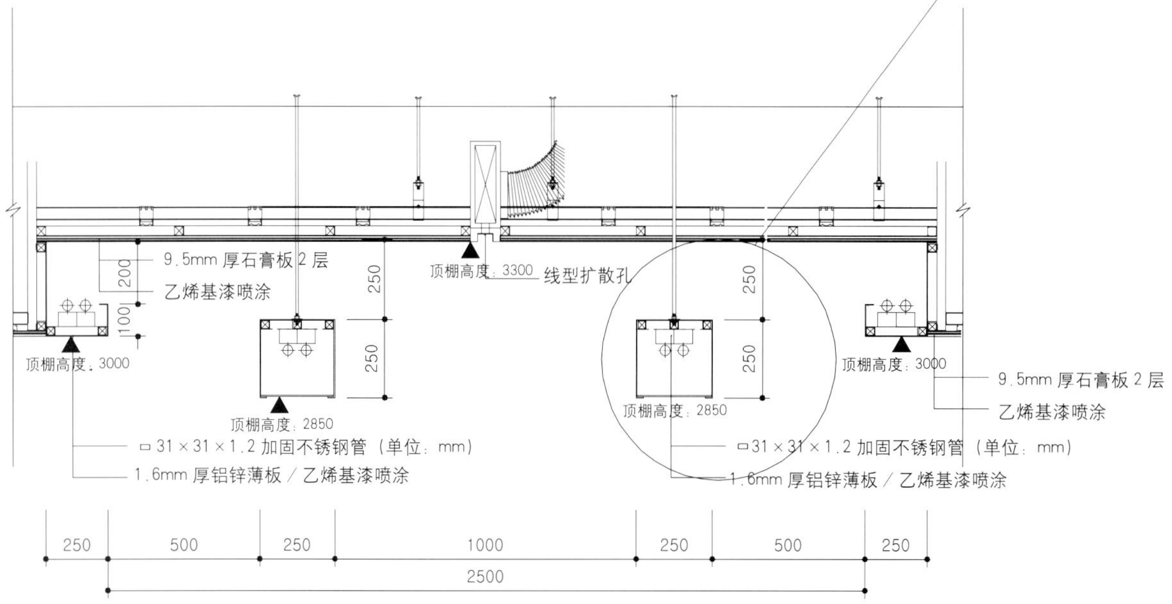

灯箱盒一剖面图和细部图

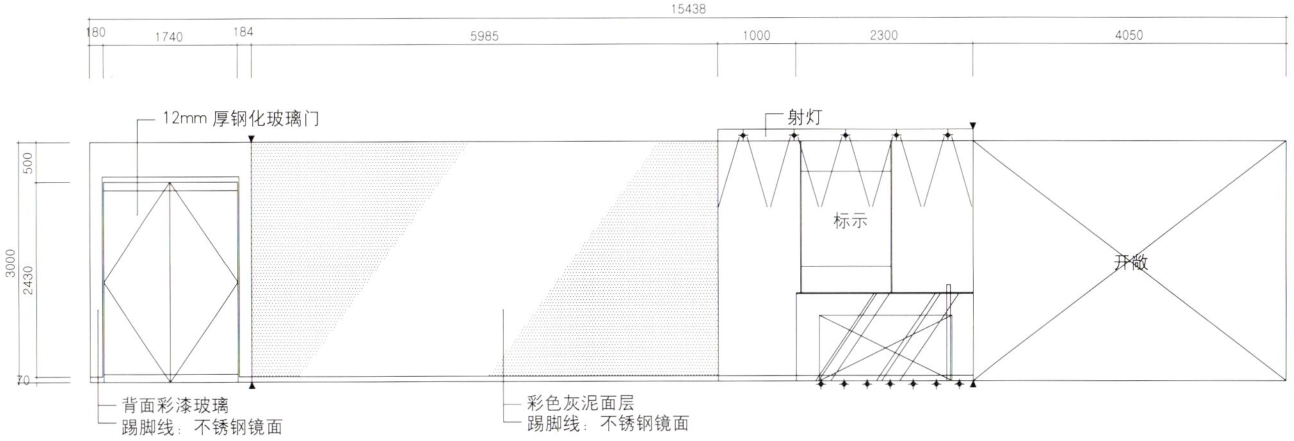

餐厅一立面图 A

设计：设计秀国际公司 朴志荣 李显庆 (02) 546—0220 | **施工**：设计秀国际公司 韩光宪 | **地址**：大田市西区东山洞 1455 番地 | **面积**：264m² | **装修**：地面／釉面砖，木地板，瓷性瓷砖 墙体／晨石，树脂，木纹薄木板，进口布料，丝织墙纸 顶棚／树脂喷涂

Design: DESIGNSOO International · Park, Ji young / Lee, Hyung kyung I **Construction:** DESIGNSOO International · Han, Kyang hyun I Built Area: 264m² I **Finish:** Floor / Polishing Tile, Wood, P-Tile, Ceramic Tile Wall / Morning Stone, V.P, Veneer Sheet, Imported Fabric, Silk Wallpaper Ceiling / V.P

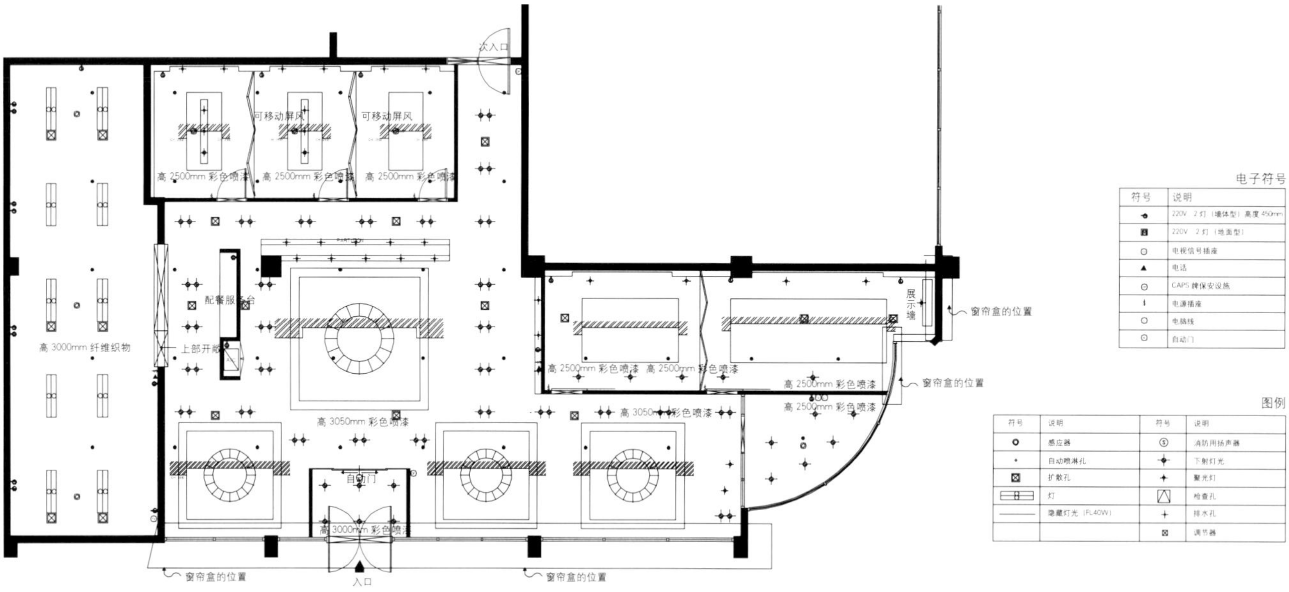

顶棚平面图

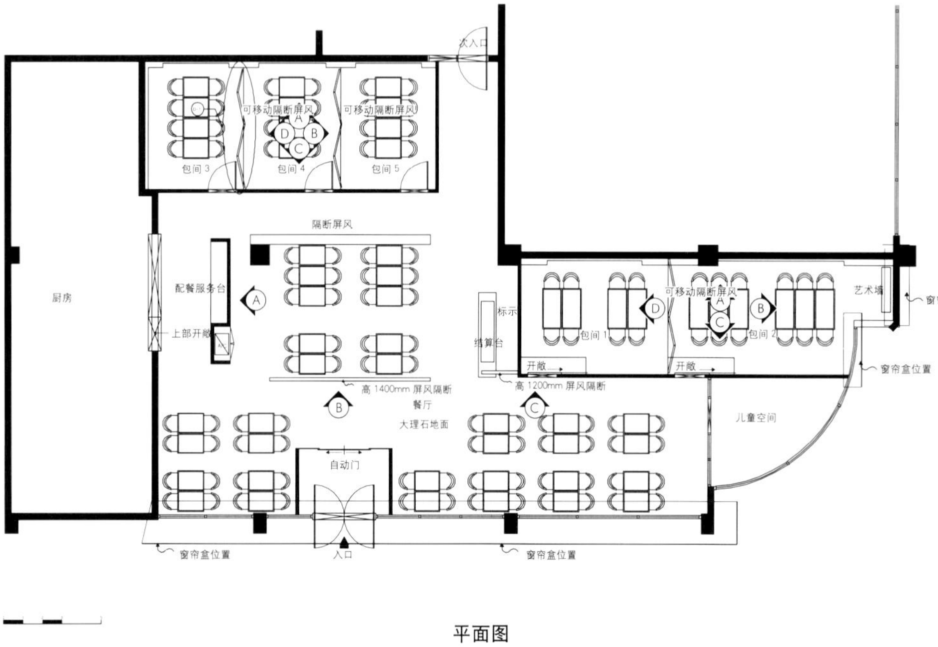

平面图

“景同情异”。铉京饭店的设计理念很独特，虽然是中国式的传统餐馆，但却没有采用传统的装饰手法。设计师希望营造一种气氛，客人踏入该饭店以后能够先坐下来休息一下，喝杯热茶。与其他传统的铉京饭店连锁中餐馆不同，该饭店卫生整洁，进餐的客人之间也没有那种家族式的长幼辈之间明显的礼仪隔阂。

铉京饭店有门厅、一号包间、二号包间三种空间形式，每一种都能代表各自不同的空间感受。门厅全部采用白色和红色作为基调底色。同时为了保持各种空间之间的均衡，在门厅的几何中心布置了装饰墙和圆形灯箱。厨房配餐台和门厅空间之间的分界线用红色的屏风遮挡，给人留下了深刻的印象。

一号包间用移动式带轮式屏风将空间分为三部分，这个包间最大的优点是可以灵活分割兼作会议室使用。经过业主和设计师之间的沟通，这里设计成了为当今繁忙的现代人提供的“时间”礼物。房间如果完全封闭，噪声将会很大，于是通过移动滑轮式屏风进行部分隔声。各种空间都采用不同的墙面，加上顶棚上的红色灯箱。这样无论开放还是封闭的空间均能给人留下深刻的印象。

二号包间面对着大街，与大街上行人的视觉交流成了它的亮点。这里的室内外分隔没有采用通明玻璃而是采用了罗马帘(Roman Shade)，与墙上的织物纤维、顶棚上的悬挂的圆形吊灯一起把整个空间装扮得更加典雅舒适。

■ 餐厅

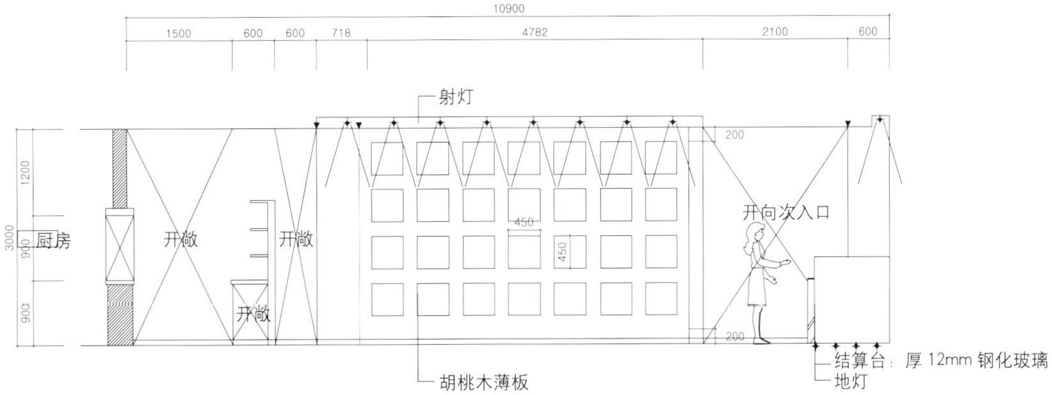
10900
1500
600
600
718
4782
2100
600
射灯
200
1200
3000
900
900
厨房
开敞
开敞
开敞
450
450
开向次入口
200
结算台：厚 12mm 钢化玻璃
地灯
胡桃木薄板

立面图 B

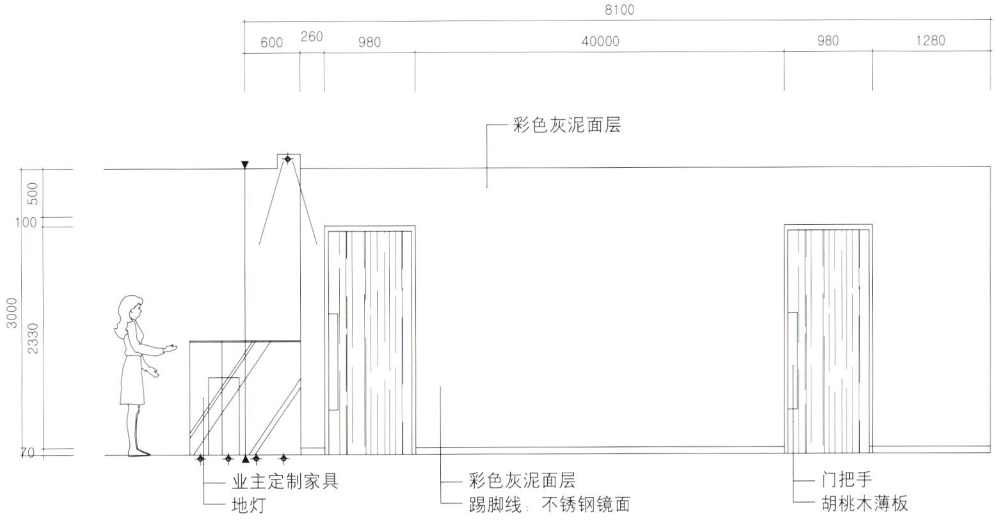
8100
600
260
980
40000
980
1280
彩色灰泥面层
500
100
3000
2330
70
业主定制家具
地灯
彩色灰泥面层
踢脚线：不锈钢镜面
门把手
胡桃木薄板

立面图 C

■ 包间1、2

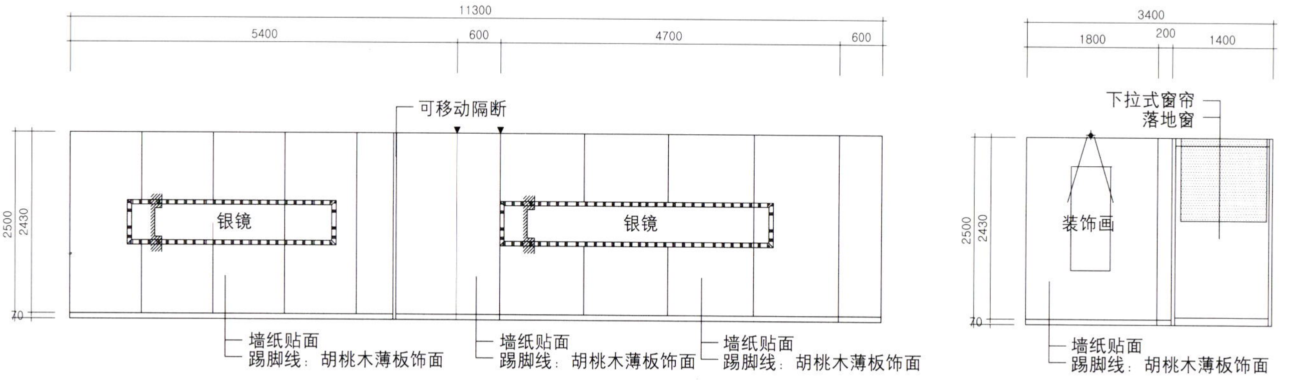

立面图 A、B

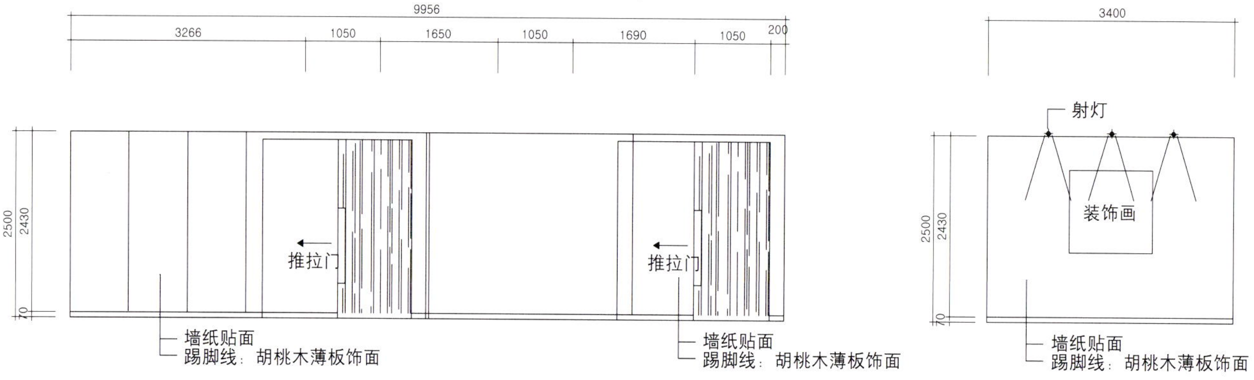

立面图 C、D

■ 包间3、4、5

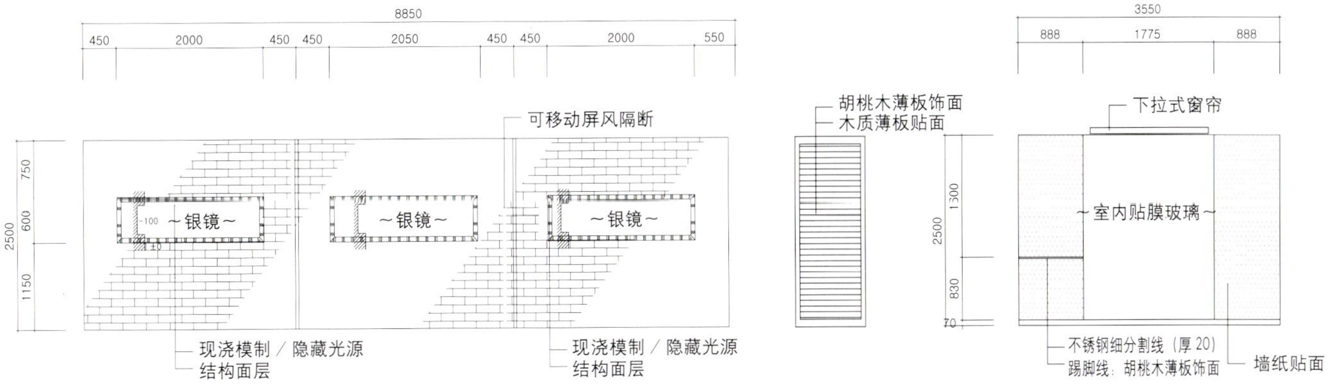

立面图 A、B

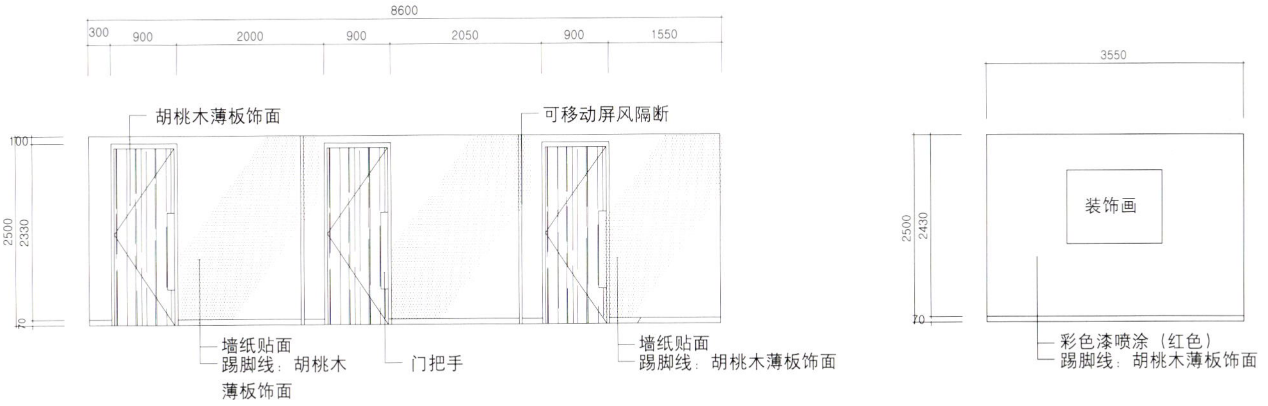

立面图 C、D

'Same place with different feelings'. A space excluding the Chinese traditional feel, but where one eases oneself and has a cup of tea. This is a concept of Hyun kyung. Unlike the previous branches, this place does not follow the looks of a Chinese restaurant and transformes itself into a new space with a neat and refined image that could break down the generation gap.

This Hyun kyung comprises a hall, room 1 and room 2, each space with different feels. Main red and white colors were used in the hall, the center of which occupied with a display shelf and round lighting box as the centroid. A red partition sets a boundary between a serve table and hall to brighten the image.

In Room 1, a moving partition divides the space into three.

The biggest strong point is that it was also designed to have a function of a conference room. This, the result of a tuning between the client and designer, would provide the busy modern people with a big gift named time. A moving partition was installed to solve a noise problem in case the space is closed. Each space creates a unique atmosphere, which changes according to its state of being open and closed, by the unique wall image and red lighting box.

Room 2 is a communicating space for people inside and outside. It is the responsibility of a Roman shade instead of a glass to separate the inside from the outside. Fabric on the wall and a round illumination of the ceiling add coziness.

■ 包间3、4、5—灯箱盒

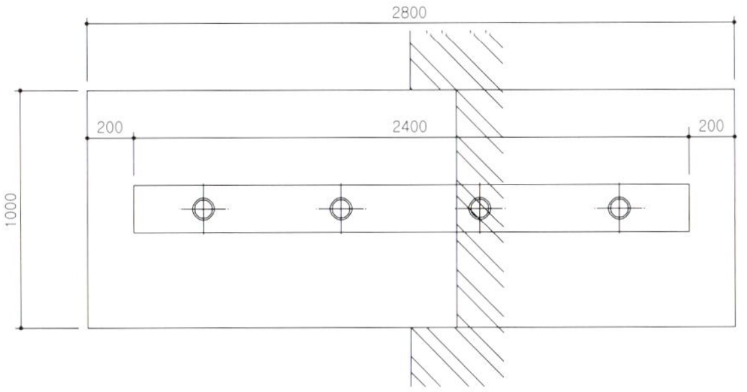

局部平面图

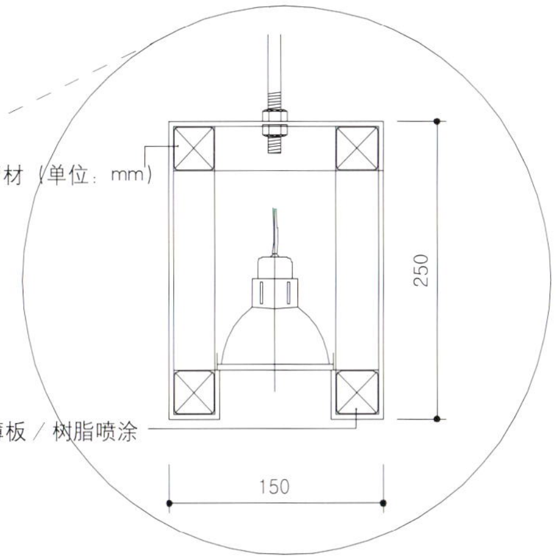

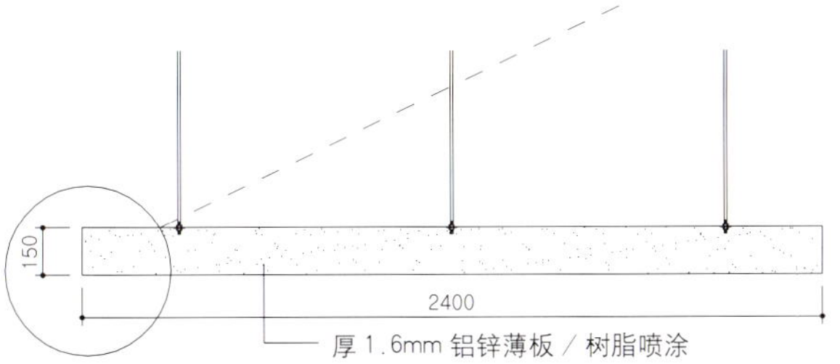

立面和细部图

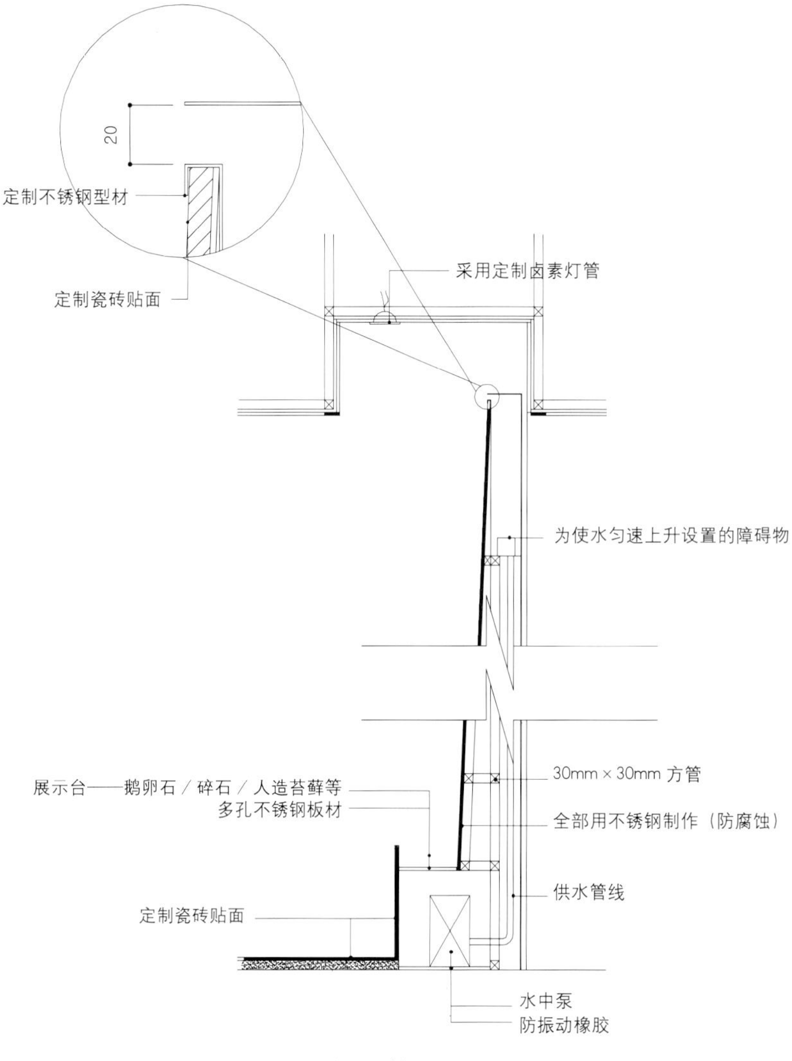

入口—剖面图 A

词典上查到的 Well-Being 的涵义是“福祉 - 安宁 - 幸福”等意思。最近，这一说法有了进一步的引申：就是身心相互配合，探讨多彩人生的新生活方式或者文化体验也叫 Well-Being。同样，那些能按照自己的生活方式保持身心健康，并由此体验到幸福感的人群，则被称之为 Well-Being 一族。这几年随着 Well-Being、健身迷、晨练等生活方式的流行，与此相关的时髦流行词也多了起来。但与此同时，禽流感、疯牛病等疾病以及劣质冻水饺等有害食品也趁机出现在人们的日常生活中，人们不禁要问“今天到底应该吃什么”，这就是关心自己“健康”的反映。Well-Being 一族不吃猪肉、不喝碳酸饮料，也不使用化学调味品，他们喜欢吃原汁原味的海鲜和绿色蔬菜。

因此，真正的 Well-Being 一族永远保持着健康的身心，不再使用化工产品，而积极投入大自然怀抱，与自然为伍，他们是绿色食品的倡导者。

“健”就是伴随着 Well-Being 一族的出现，应运而生的韩式餐馆。设计师从健康设计的原理出发，研究了东方哲学的所谓风水观以及自然要素中的阴阳五行概念。

“万物来自于光明与黑暗，土、石、火、水、木……”再也没有比用自然材质装扮空间更合适的了。

翻开菜谱看到的首先不是传统的韩食食谱，而是以保健为主的健康食品。就连建筑空间也具有韩国的瓦房、日本的榻榻米、中国的汉字一样的特点。随心所欲、快乐安详，饭菜色、香、味俱佳，还能经历别样风格的空间感受，这是多么有趣的就餐时光呀！

设计： 亚工设计研究所　金武贤（02）421–5797 | **施工：** 室内艺术设计　金中宪 | **地址：** 首尔市江南区新舍洞 524–12 道京大厦一层 | **面积：** 194m² | **装修：** 地面／瓷砖、榻榻米　墙面／乳胶漆喷涂、韩宣纸壁纸、瓷砖、砖、瓦、黑镜　顶棚／乳胶漆喷涂、防腐木材、金属、竹木

Design: Agongdesign Associates · Kim, Wook seon | **Construction:** Interior ART · Kim, Jung hyeon | Built Area: 194m² | **Finish:** Floor / Tile, Tatami Mat Wall / Painting, Koreanpaper Wallpaper, Tile, Brick, Roofing Tile, Black Mirror Ceiling / Painting, Antiseptic Wood, Steel, Bamboo Screen

■ 入口

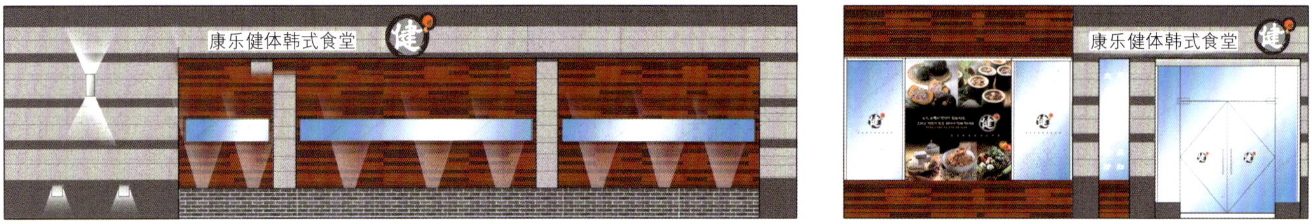

立面图 A

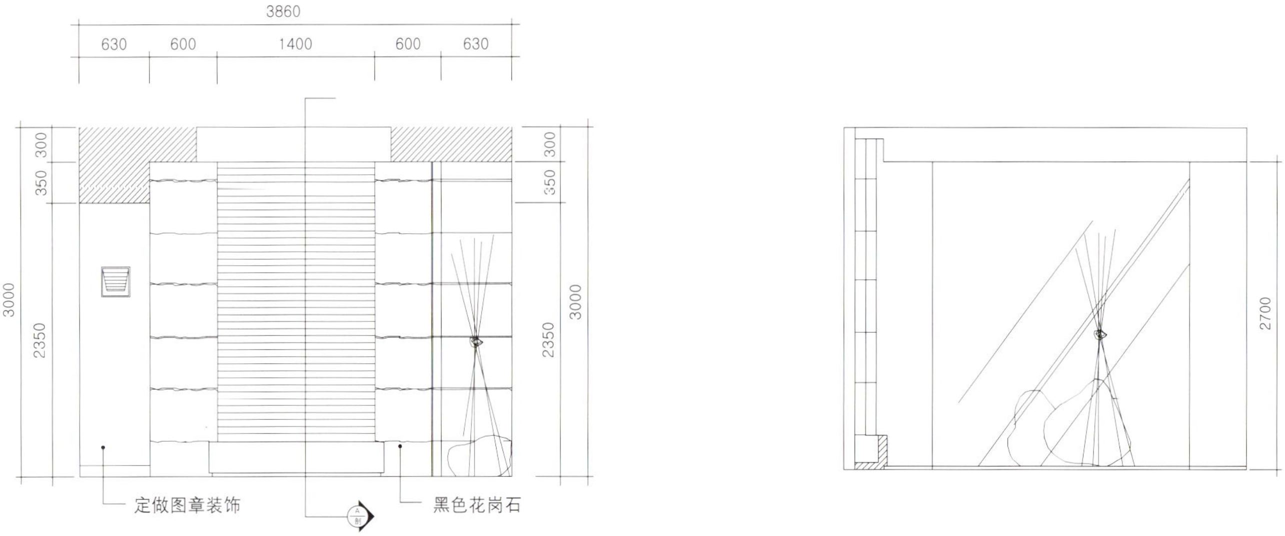

立面图 B、C

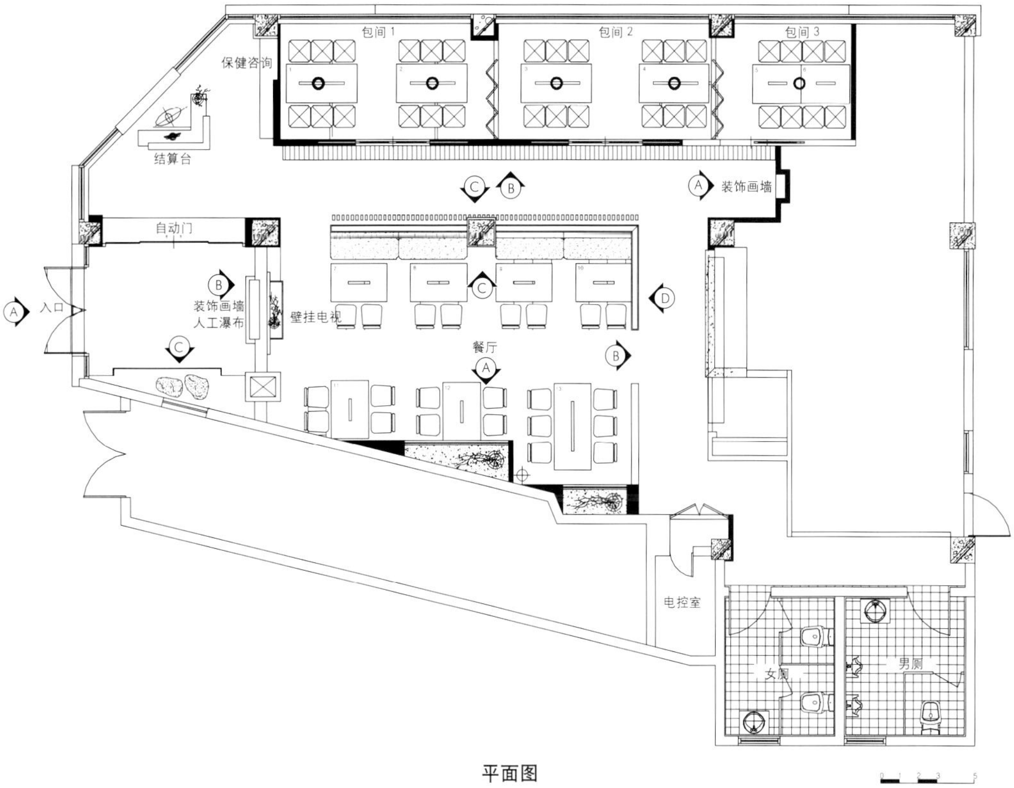

平面图

■ 餐厅

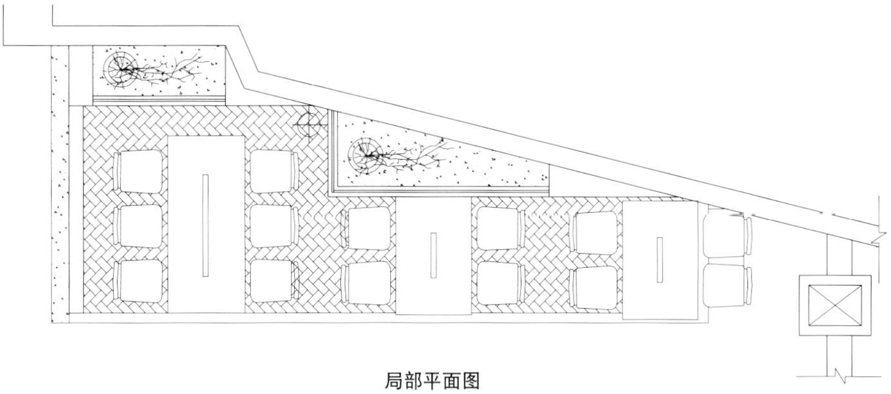

局部平面图

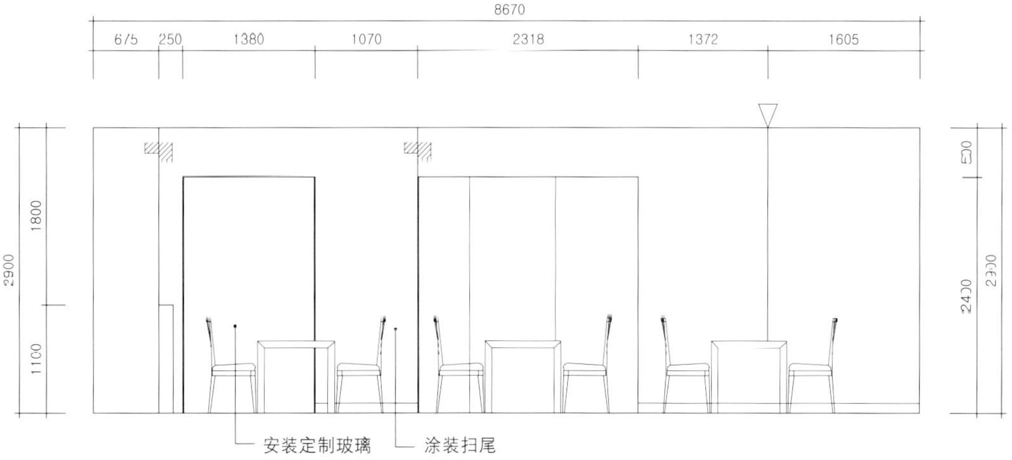

立面图 A

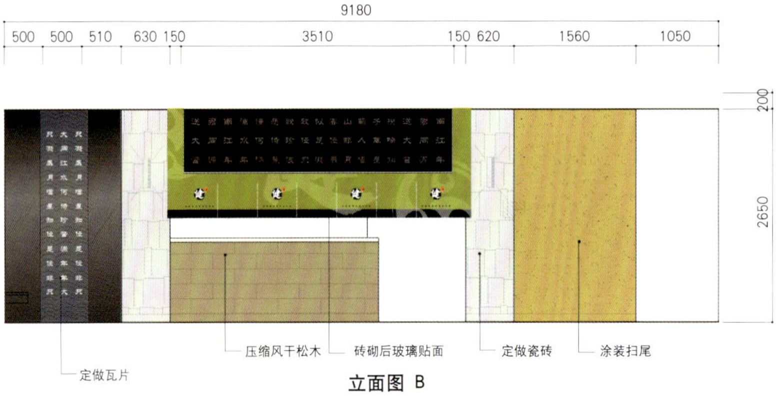

立面图 B

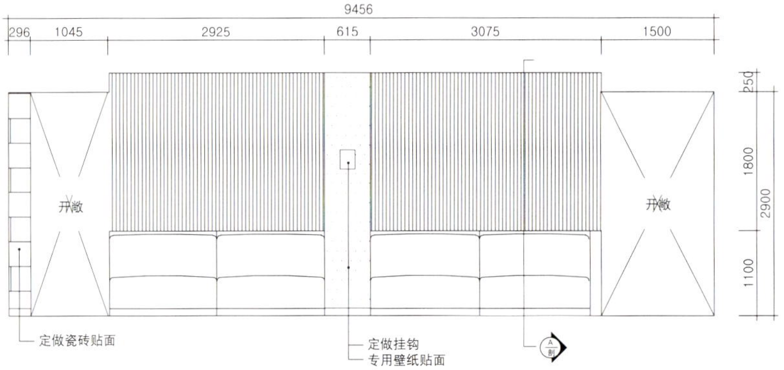

立面图 C

Dictionary defines 'Well-being' as welfare, peace and happiness. Today, it is interpreted as a new type of lifestyle or cultural code according to which one lives a rich and beautiful life through the organic combination of body and mind. 'Well-being' people are who obtains happiness from a sound body and mind maintained in their own ways. New items are pouring out one after another, in connection with Well-being, Momjjang(well-built) and Morning person syndrome. After bird flu, mad cow disease and the dumpling scandal, people pay full attention to what to eat. Well-being people eat fish and organic food instead of meat, and avoid chemical flavor and carbonated drink. The focus is "health". In my opinion, the true well-being can be only achieved when a body returns to nature and be organic, freeing itself from chemical products.

The designer of Gunn stepped the first step toward healthy design in 'nature'. It exhibits Feng-Shui theory of Eastern philosophy and the basis of nature, Yin-Yang & Five Elements. 'Fill it with the origin of all things.' Light, dark, earth, stone, fire, water and tree - these were thought as the best materials.

The menu tells us that they serve Fusion dishes, not Korean traditional food. It excellently explains the harmony among Korean tile, Japanese tatami and Chinese character. A perfect fit. Amusing. It amuses one's sight and taste. And there is also an important element differentiated from that of other restaurants. The space is made of true well-being and takes pride in itself as a background role for its products.

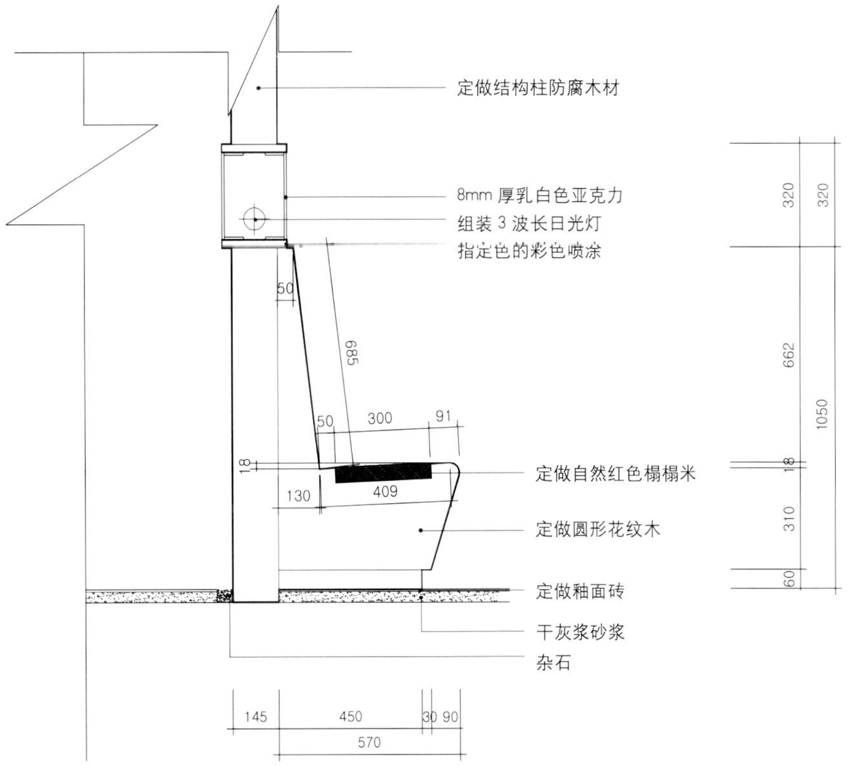

剖面图 A

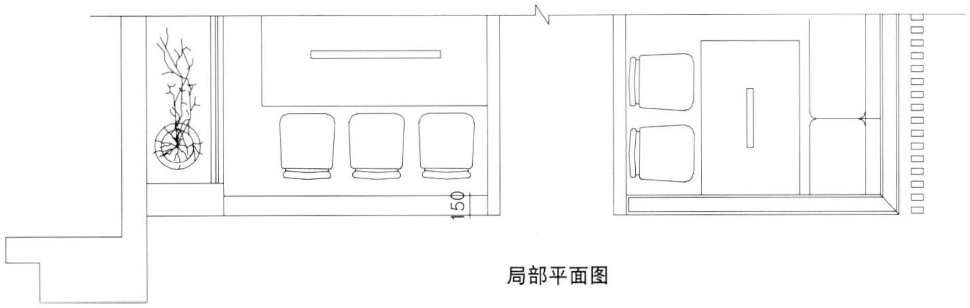

局部平面图

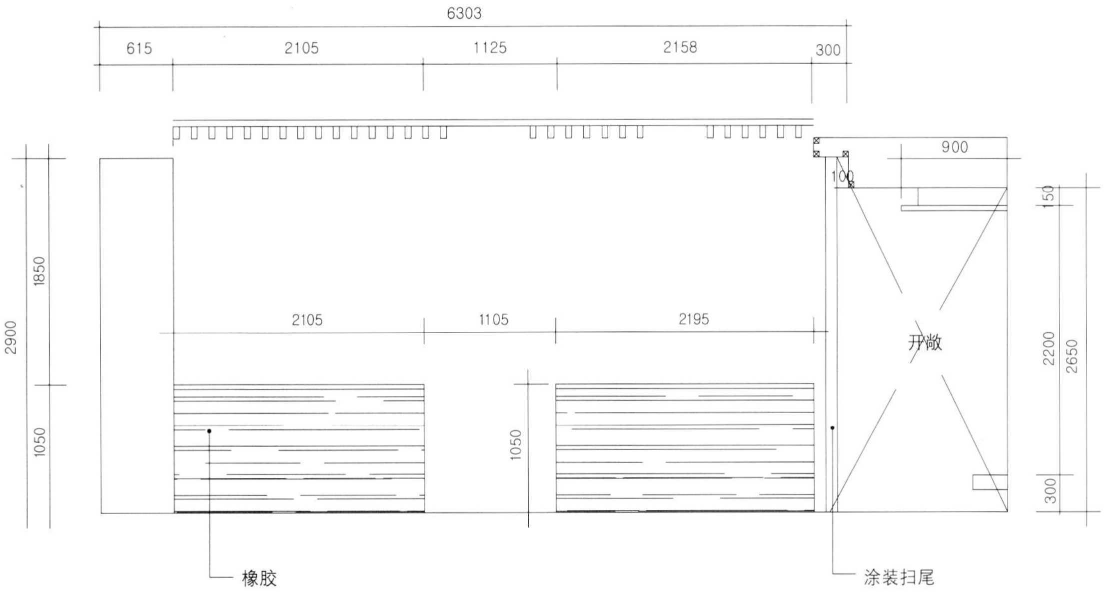

立面图 D

■ 走廊

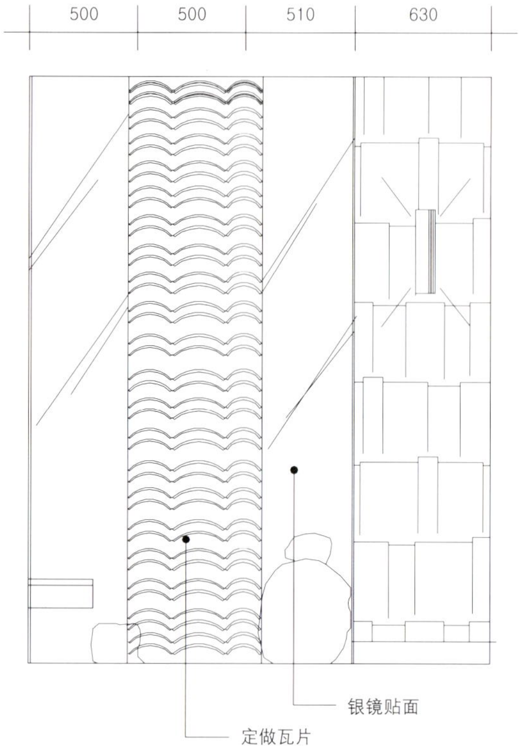

装饰画墙—立面图 A

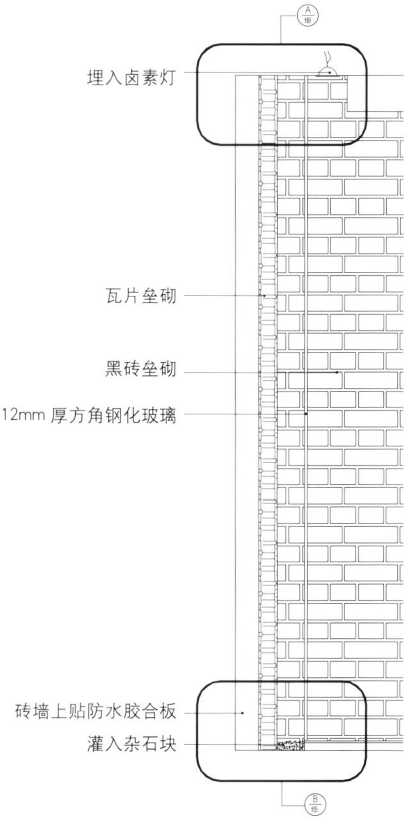

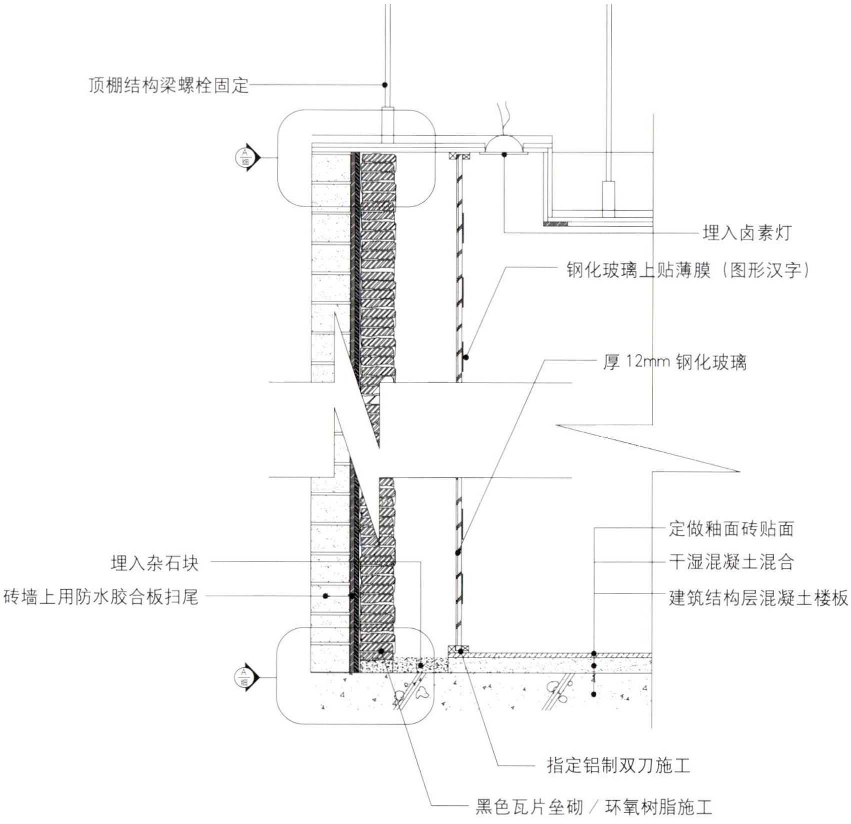

剖面和细部图 A、B

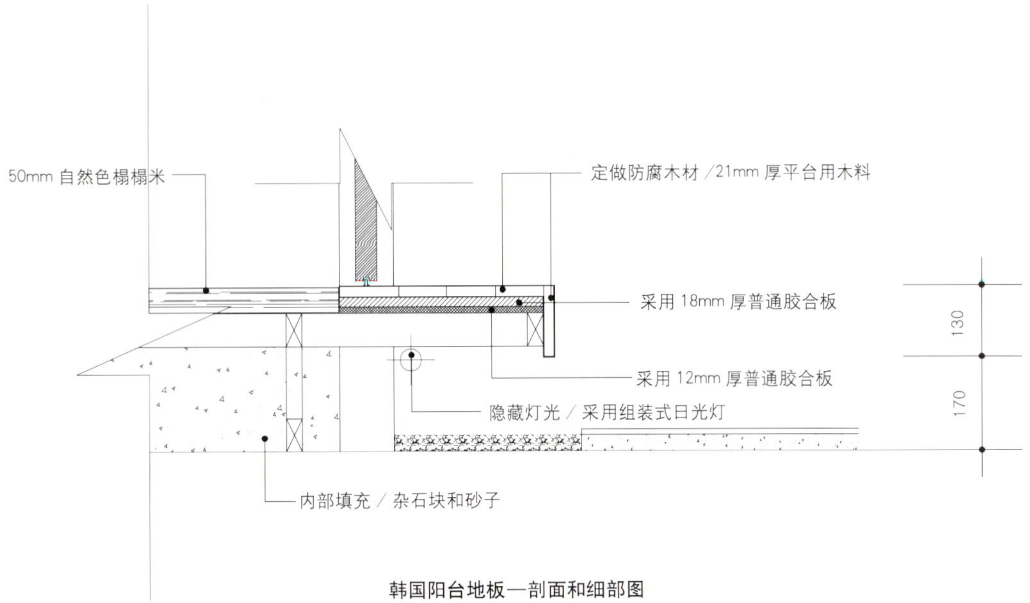

韩国阳台地板—剖面和细部图

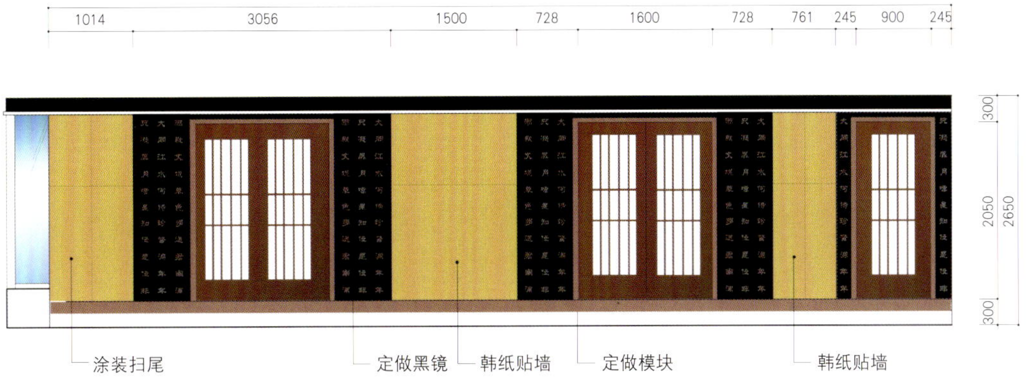

立面图 B

■ 走廊

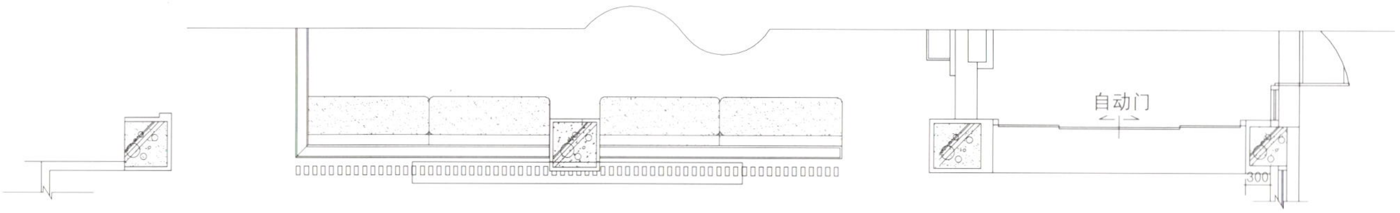

局部平面图

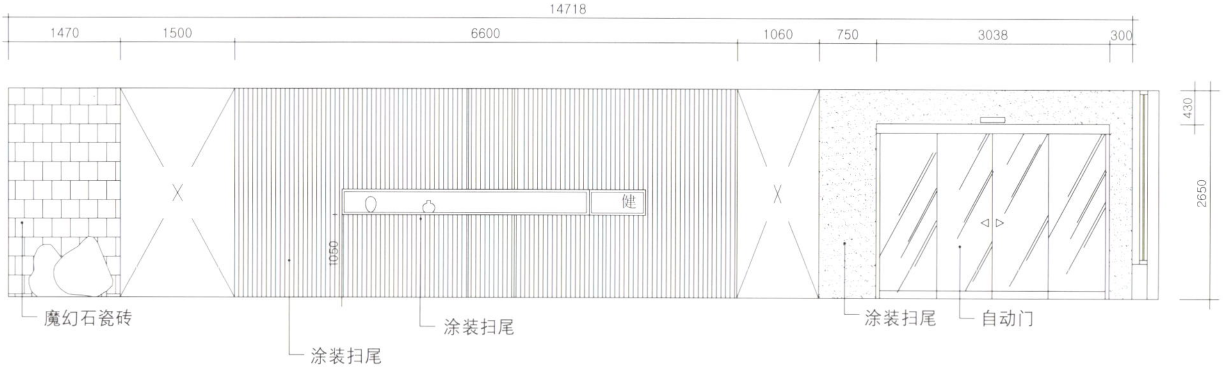

立面图 C

造型墙

■ 包间

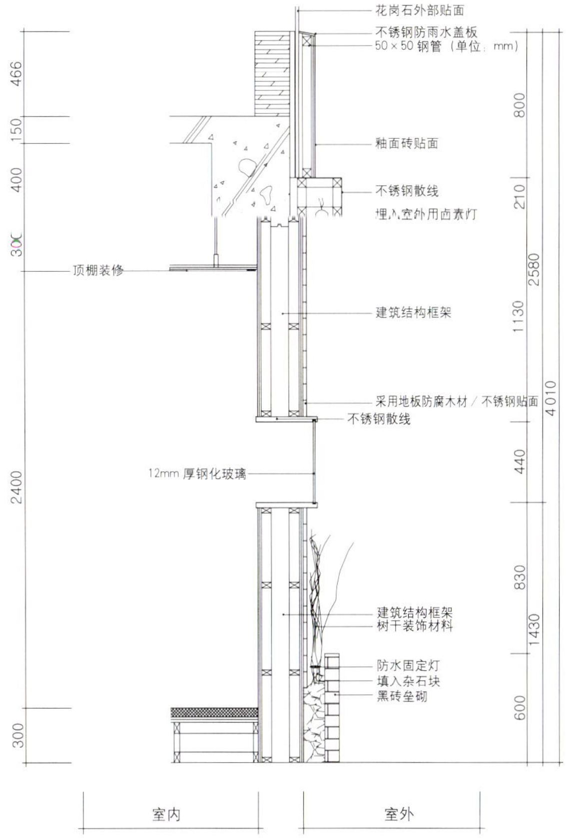

剖面图 A'

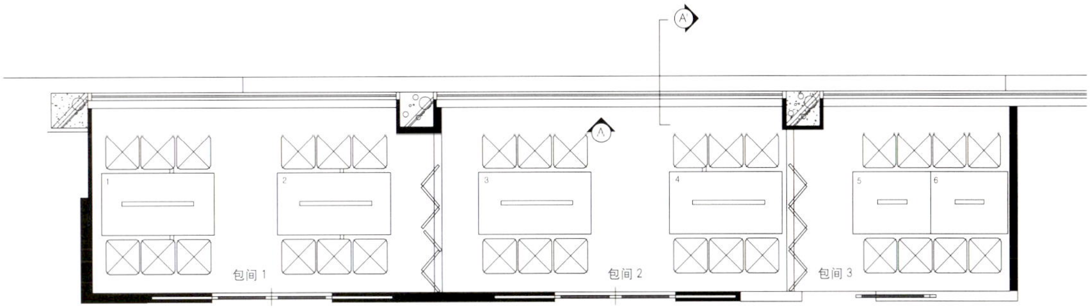

局部平面图

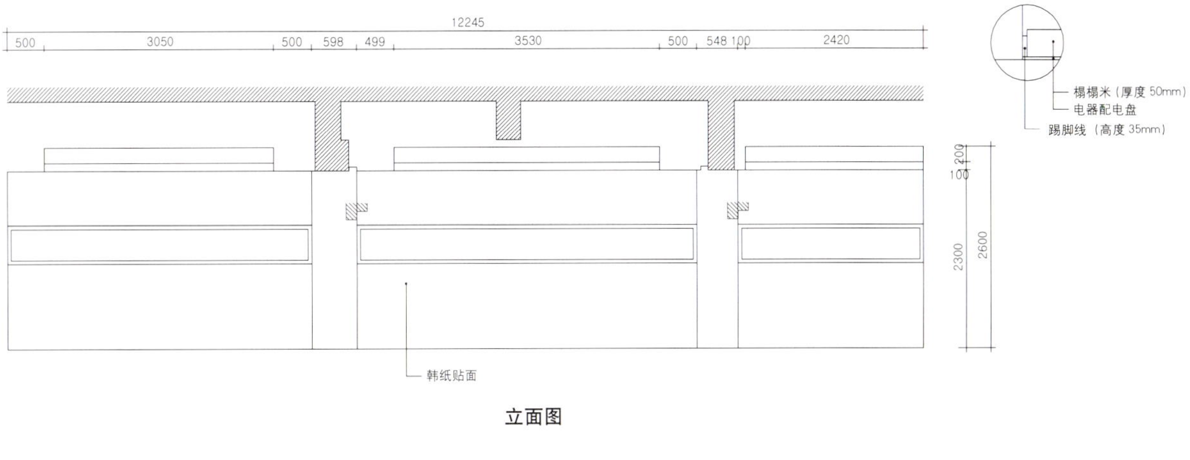

立面图

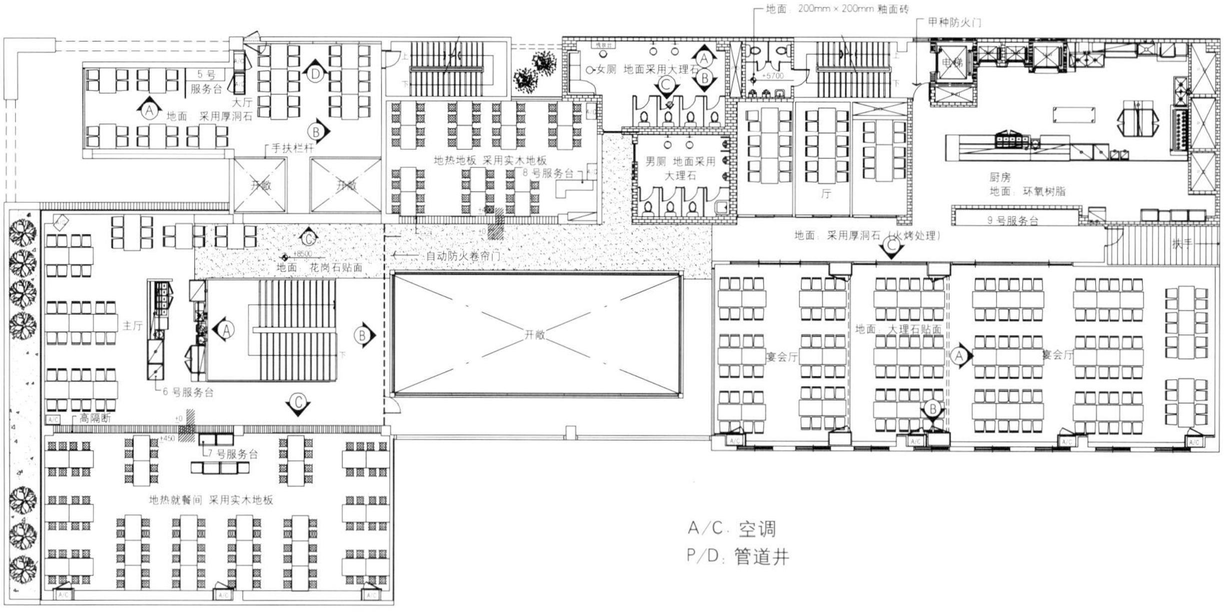
地面 200mm×200mm 釉面砖
甲种防火门
5号服务台
大厅
地面 采用厚洞石
手扶栏杆
开敞
女厕 地面采用大理石
男厕 地面采用大理石
地热地板 采用实木地板
8号服务台
厨房
地面 环氧树脂
9号服务台
地面 采用厚洞石（火烤处理）
扶手
地面 花岗石贴面
自动防火卷帘门
主厅
6号服务台
高隔断
7号服务台
地热就餐间 采用实木地板
宴会厅
地面 大理石贴面
A/C：空调
P/D：管道井

三层平面图

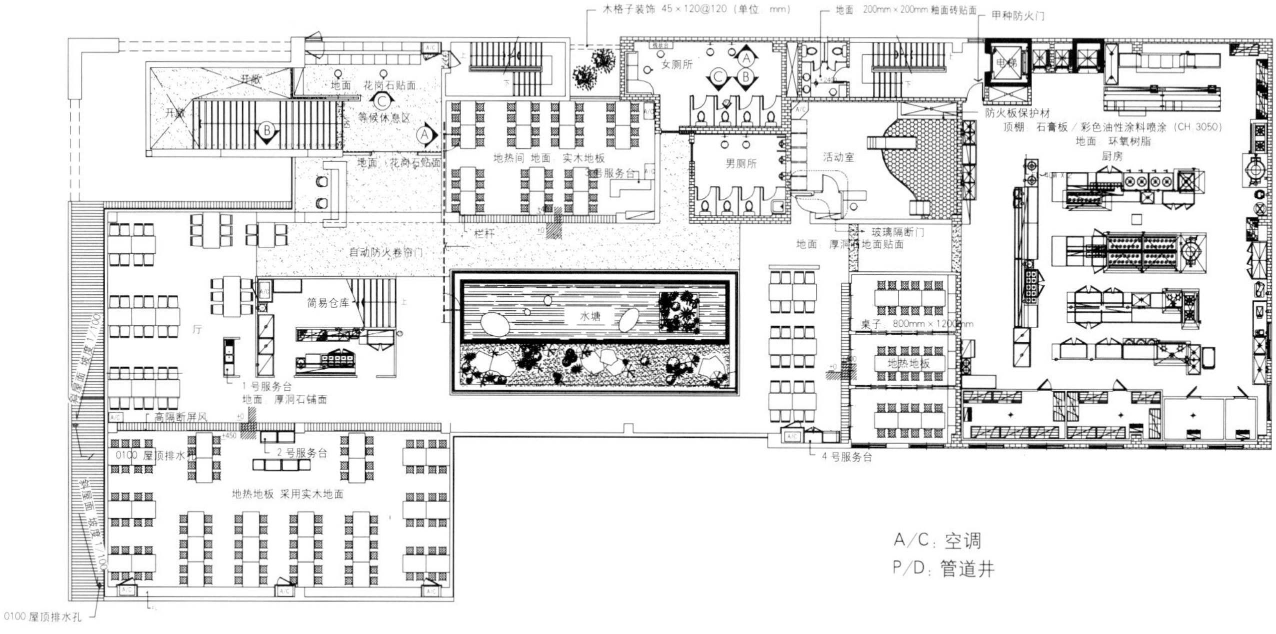
木格子装饰 45×120@120（单位：mm）
地面 200mm×200mm 釉面砖贴面
甲种防火门
开敞
地面 花岗石贴面
等候休息区
女厕所
男厕所
地热间 地面 实木地板
活动室
防火板保护材
顶棚 石膏板/彩色油性涂料喷涂（CH 3050）
地面 环氧树脂
厨房
栏杆
玻璃隔断门
地面 厚洞石地面贴面
自动防火卷帘门
简易仓库
厅
水塘
桌子 800mm×1200mm
地热地板
1号服务台
地面 厚洞石铺面
高隔断屏风
0100 屋顶排水孔
2号服务台
地热地板 采用实木地面
4号服务台
0100 屋顶排水孔
A/C：空调
P/D：管道井

二层平面图

企划设计：国民大学室内设计专业 金个川教授 | **设计：**易道建筑公司 金泰成 (02) 525-0761 | **施工图设计：**Four site公司 金在植 金在企 | **地址：**首尔市芦远区上溪洞1025-4| 面积：一层239m²、二层1199m²、三层1199m² | **装修：**地面 / 后洞石烧磨、木地板、外磨大理石 墙面 / 指定涂料涂装 顶棚 / 定做木格栅、多孔板

Planning Design: Kim, Kai chun Prof of Dept. of Interior Design, Kookmin Univ I **Design:** IDO Architecture Inc. · Kim, Tae seong I **Execution Design:** Four Site · Kim, Jae sik / Kim, Jae gi I **Construction:** Four Site · Kim, Jae sik / Kim, Jae gi I Built Area: 1F 239m², 2F 1199m², 3F 1199m² I **Finish:** Floor / Hudong stone Burner, Wood Flooring, Extradura Marble Wall / App. Stucco Ceiling / App. Wood Grille, Perforated Panel

■ 二层等候休息区

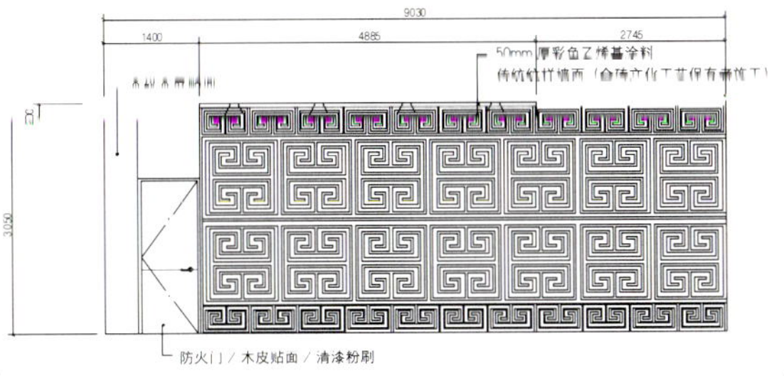

立面图 A

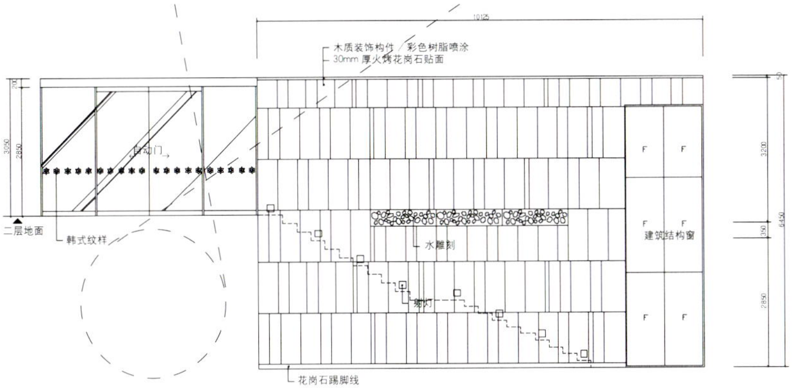

立面图 B

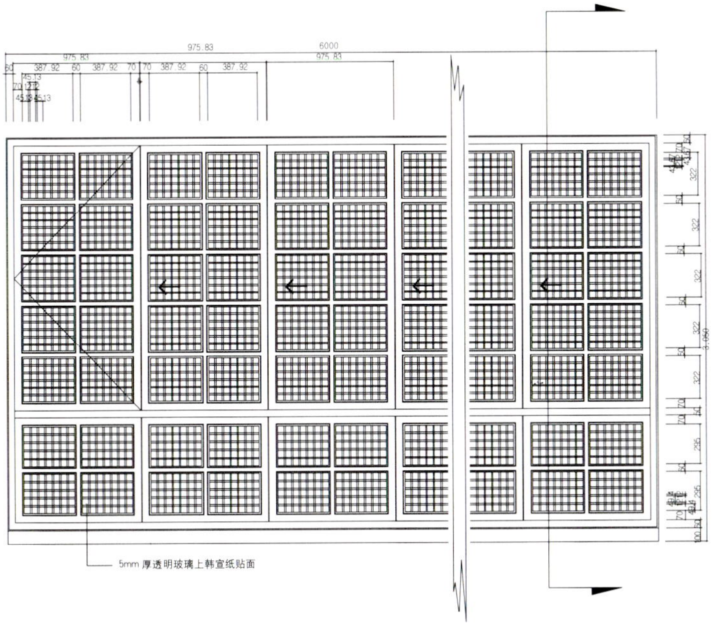

立面图 C

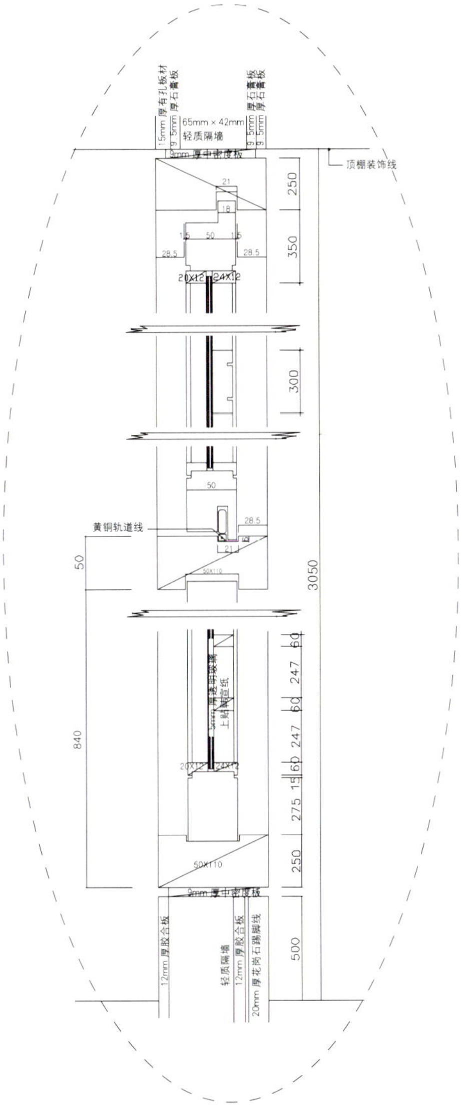

剖面图

■ 地热餐厅

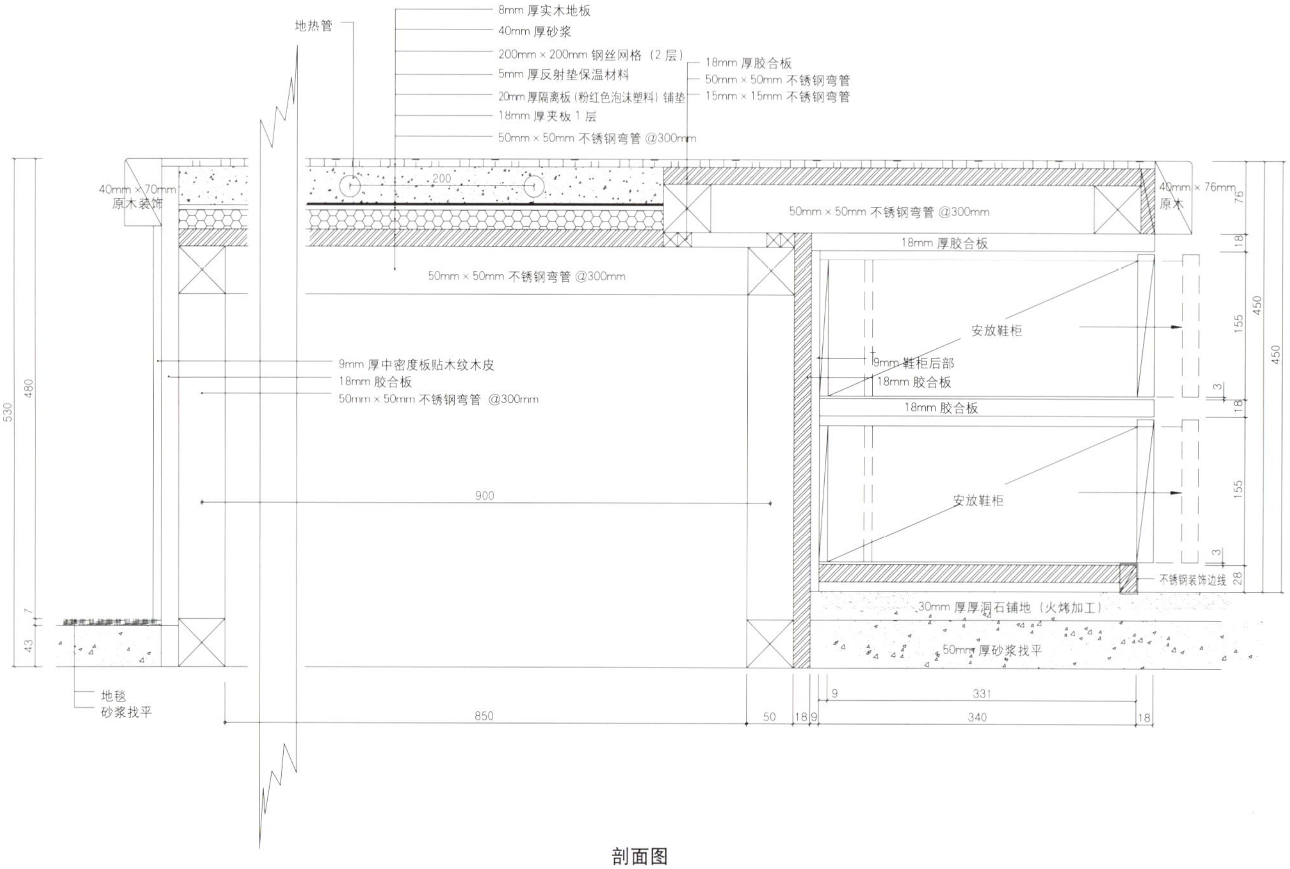

剖面图

羌羌水来烤肉店位于首尔市上溪洞，它是一座 3 层建筑，内部采用了与其他连锁店不同的布局模式。就餐者围绕着中庭落座，不仅能获得视觉愉悦，也能得到身心放松。

设计师把能感受时光流逝的天池放在了庭院中，表达了“老石深水、静木参天”的意境。

中庭水底的黑色，使水池看起来很深；周边由竹子相围，看起来挺拔高耸。即使蓝天上白云短暂飘过，映在水中的倒影也能使人感受到天地的时空变化。

与某些华而不实的餐厅不同，羌羌水来烤肉店外表不是很华丽，而内部安静休闲、典雅舒适，这正是它要传达给客人的感受。初来乍到的客人也许不会一次全部深刻了解，但通过多次身临其境的自我感受，就能顿悟到设计师那种将自然与饮食统一在一起的意图。

■ 二层水池

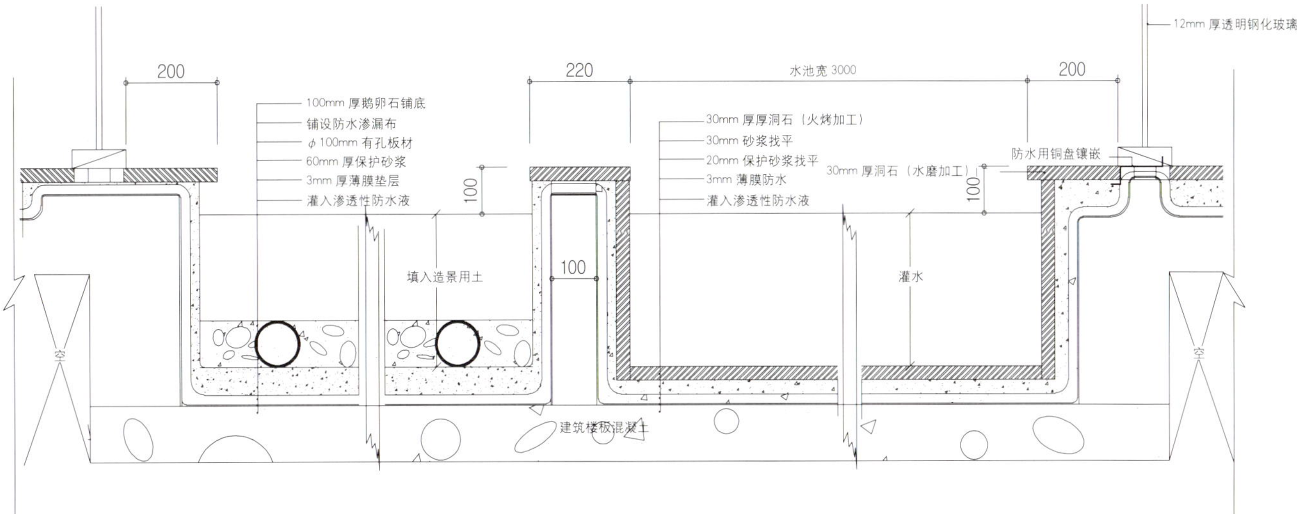

剖面和细部图

Ganggangsoolae, a three-story meat restaurant in Sanggye-dong, distinguish itself from the other commercial spaces, which usually give weight to the visual amusement; it provides a leisure atmosphere for meals and a physical/mental leeway.

The designer expressed 'No-Seok-Chim-Su, Jeogn-Mok-Cham-Cheon(Aged stone and deep water, Upright tree and blue sky)' in the courtyard that embraces the heaven and earth, from which the time flow of nature is felt.

Black stones spread on the floor give depth to a water space, and the surrounding bamboo forest stands for uprightness, with the blue sky reflected on the surface of the water suggesting the heaven and earth.

Ganggangsoolae sought a lofty(though not particularly outstanding) space that allows comfort and freedom, rather than an elegant and uncomfortable one. The designer hopes the customers would make frequent visits to experience the space little by little according to their capability and enjoy their meals in nature.

■ 宴会厅

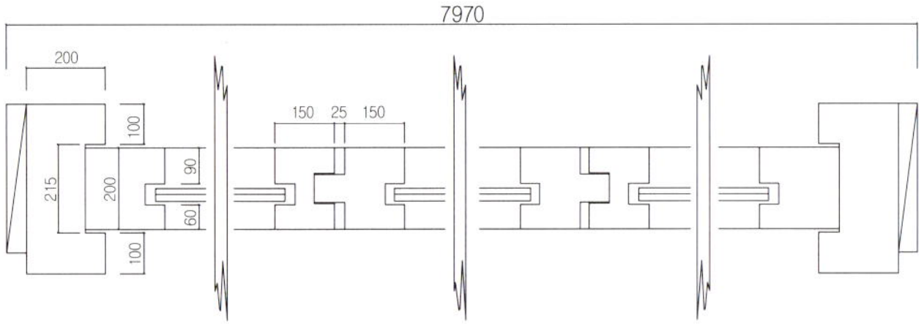

剖面图 B

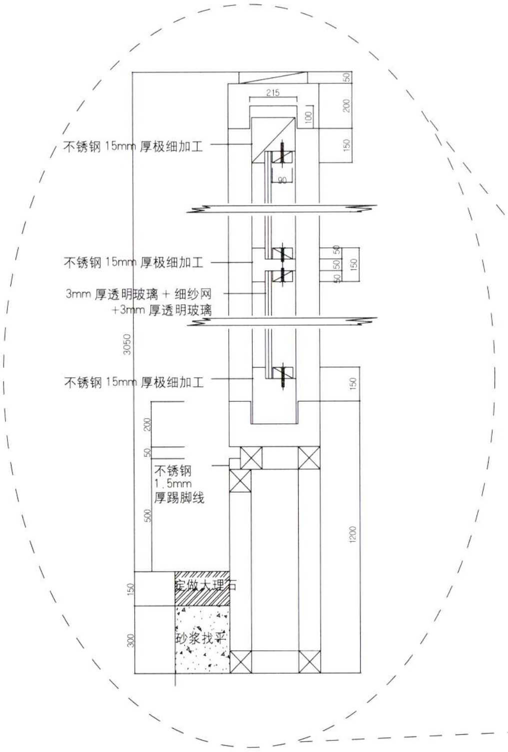

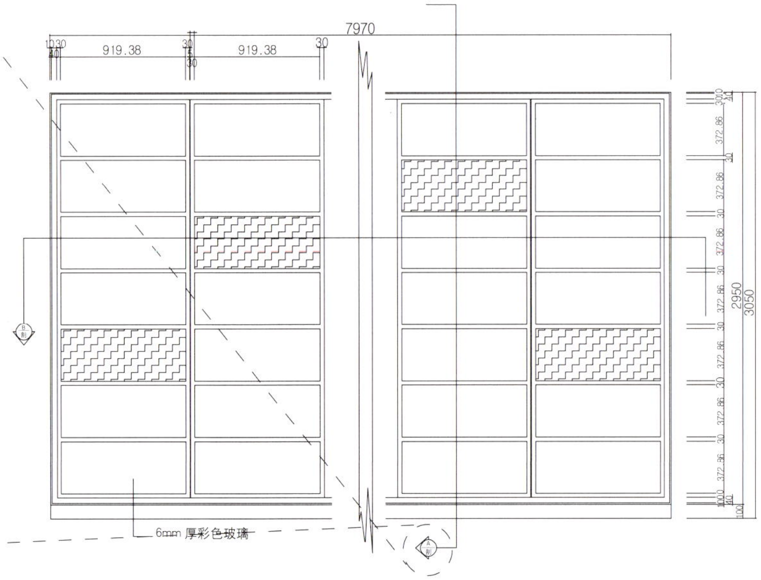

剖面图 A

立面图 A

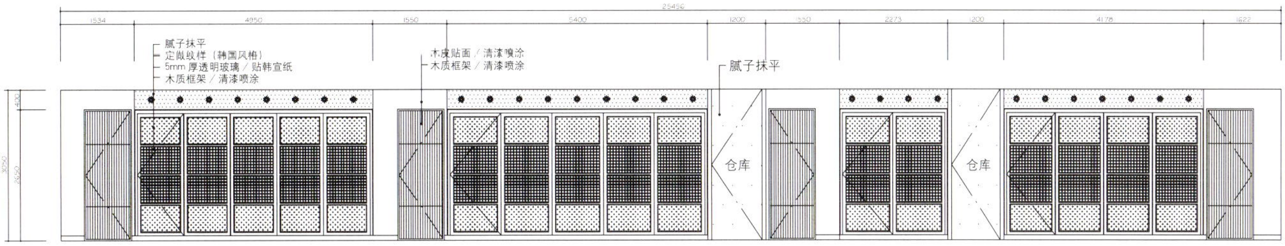
腻子抹平
定做纹样（韩国风格）
5mm 厚透明玻璃／贴韩宣纸
木质框架／清漆喷涂
木质贴面／清漆喷涂
木质框架／清漆喷涂
腻子抹平
仓库
仓库

立面图 B

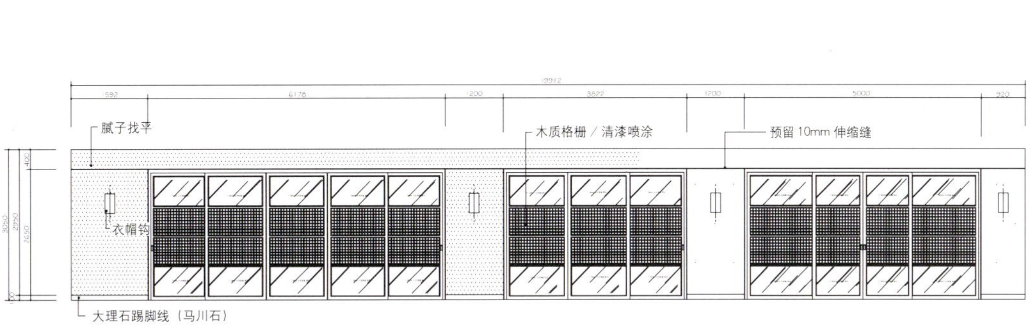
腻子找平
木质格栅／清漆喷涂
预留 10mm 伸缩缝
衣帽钩
大理石踢脚线（马川石）

立面图 C

■ 三层大厅

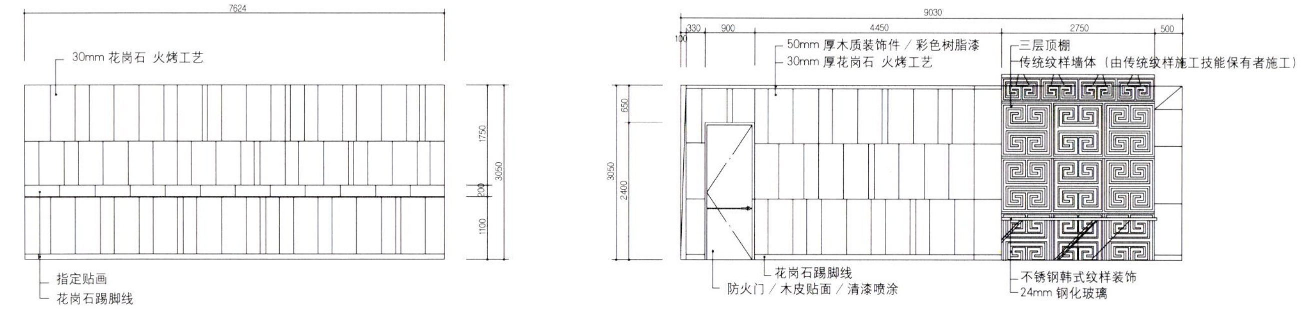

立面图 A、B

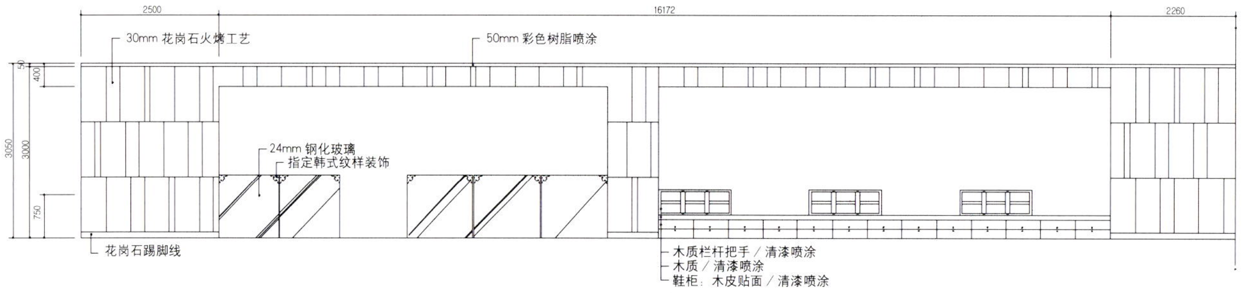

立面图 C

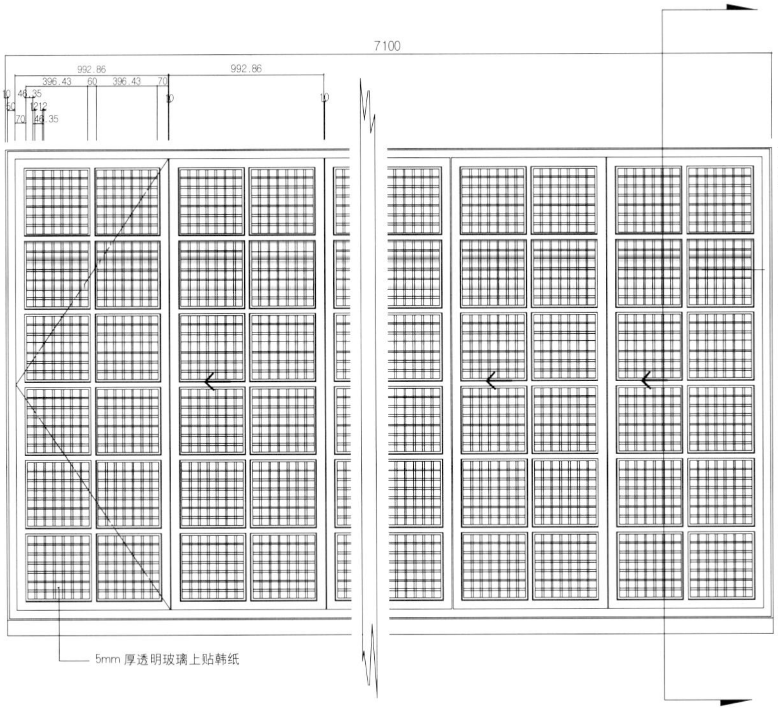

立面图 D

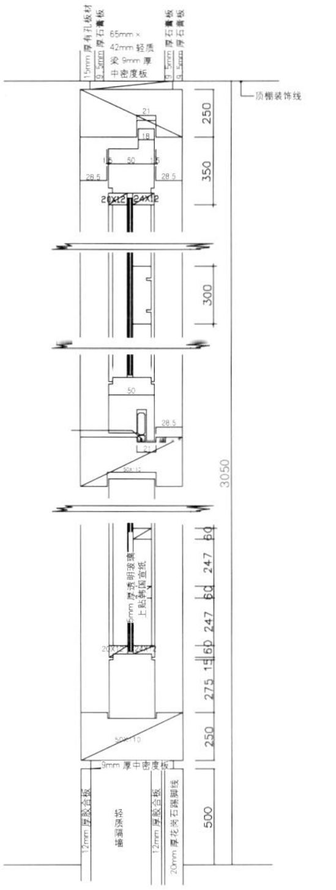

剖面图

■ 三层主厅

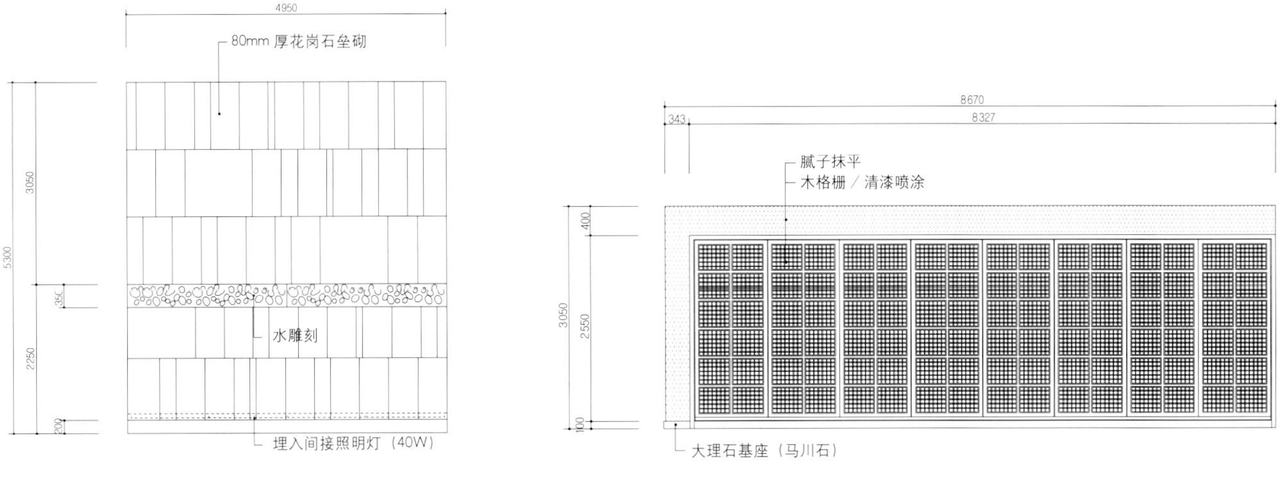

立面图 A、B

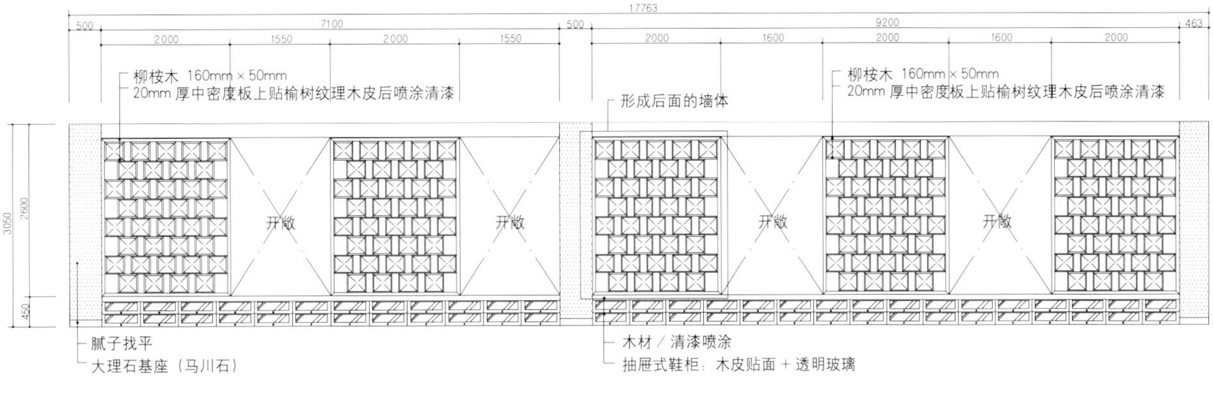

立面图 C

■ 楼梯间

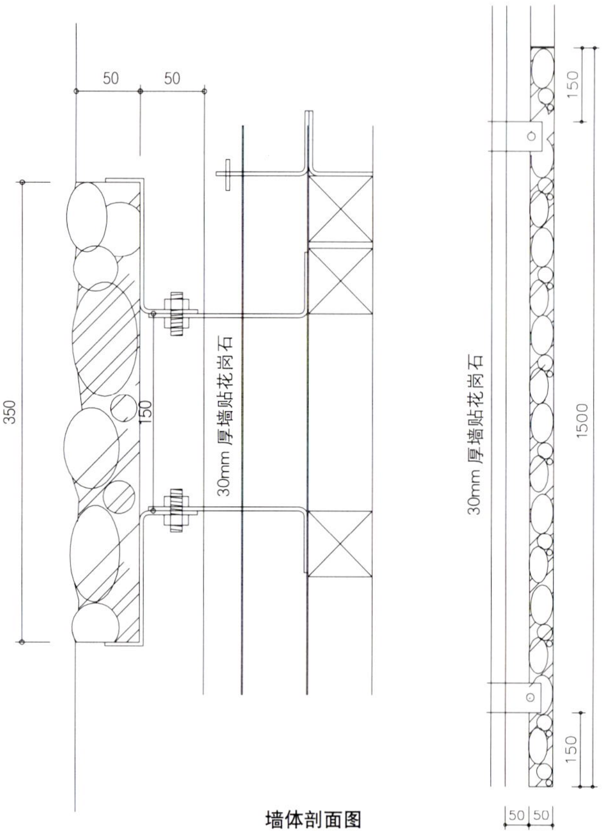

墙体剖面图

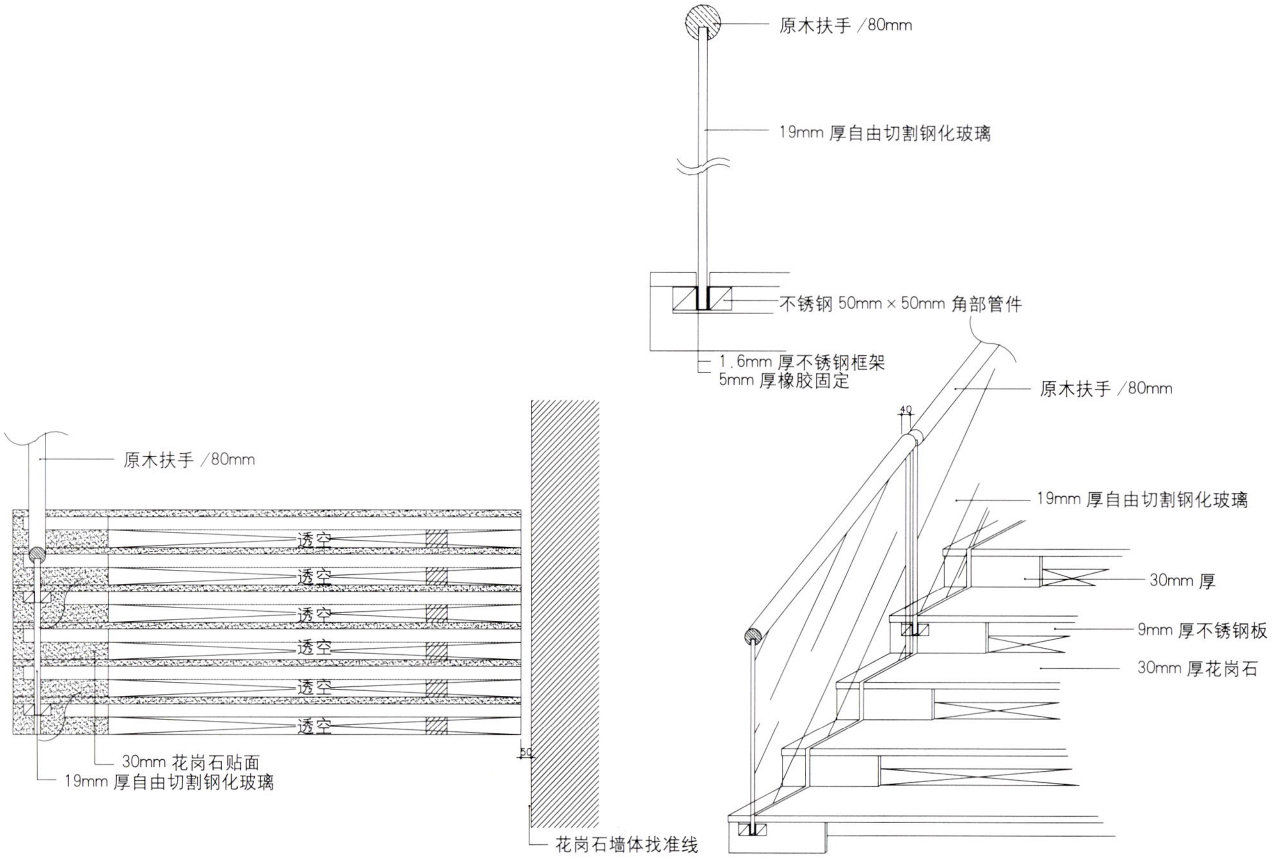

楼梯扶手一细部图

■ 墙

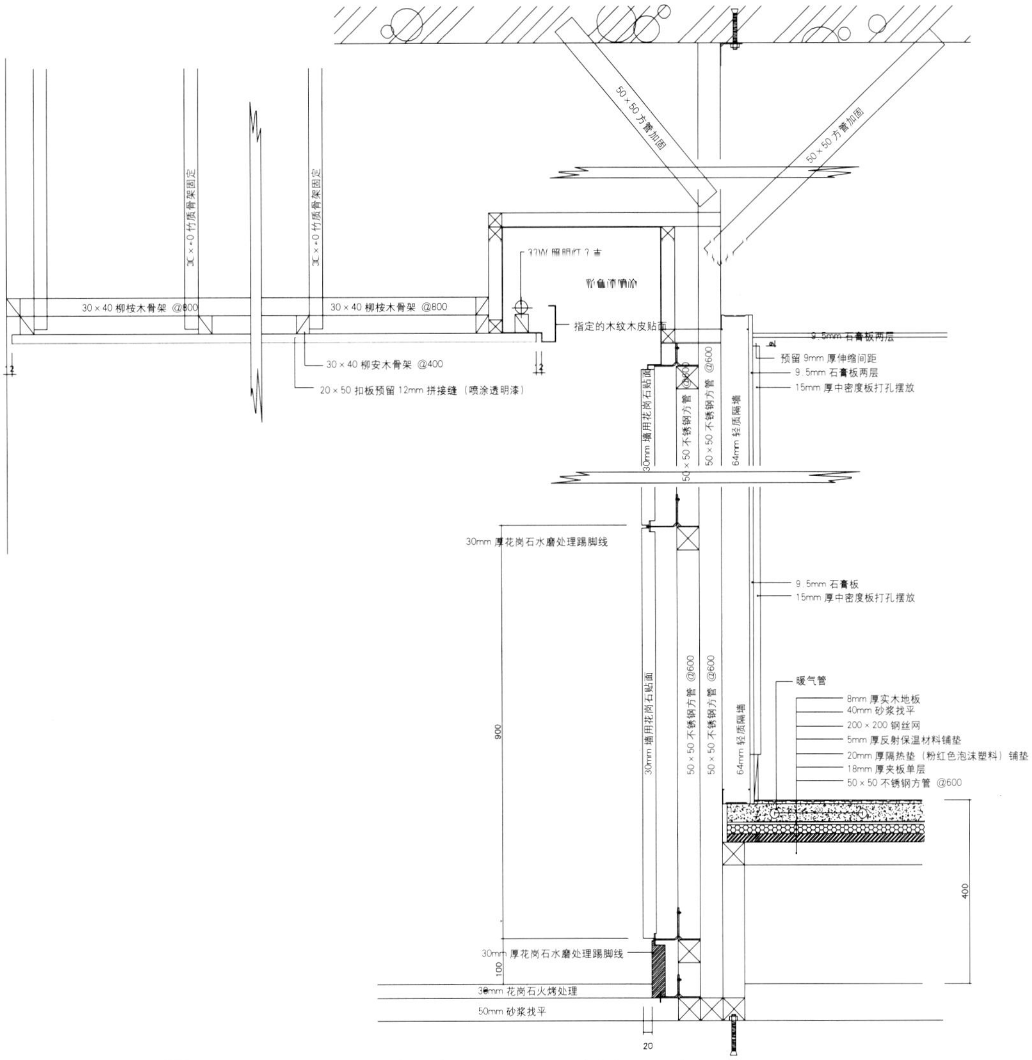

厅与走廊墙壁一剖面图（单位：mm）

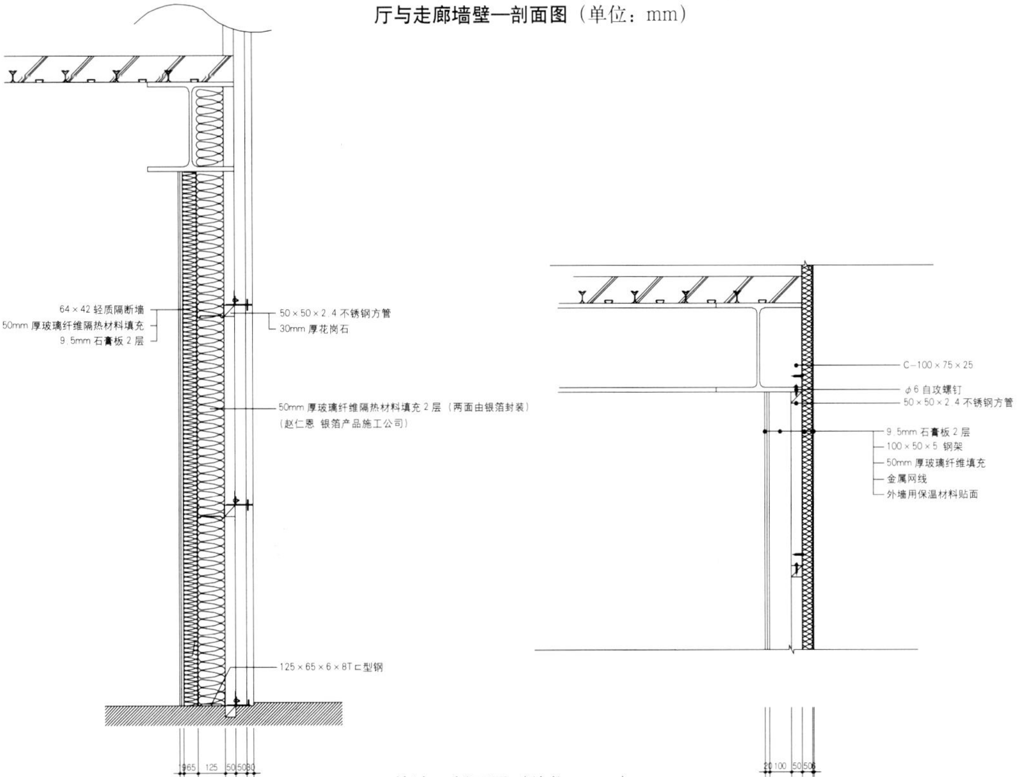

外墙一剖面图（单位：mm）

■ 女厕所

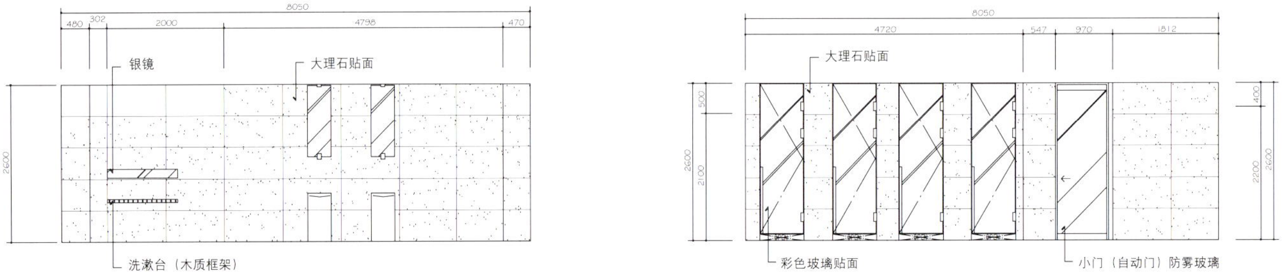

立面图 A、B

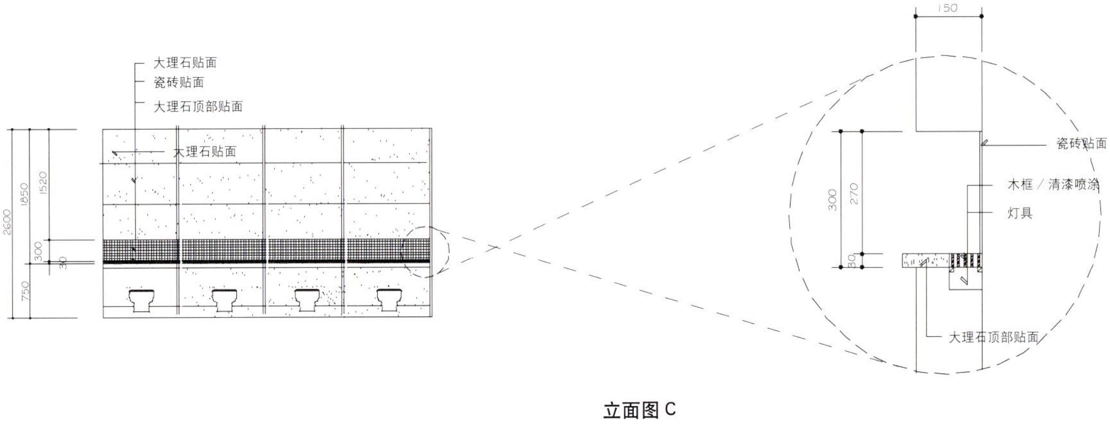

立面图 C

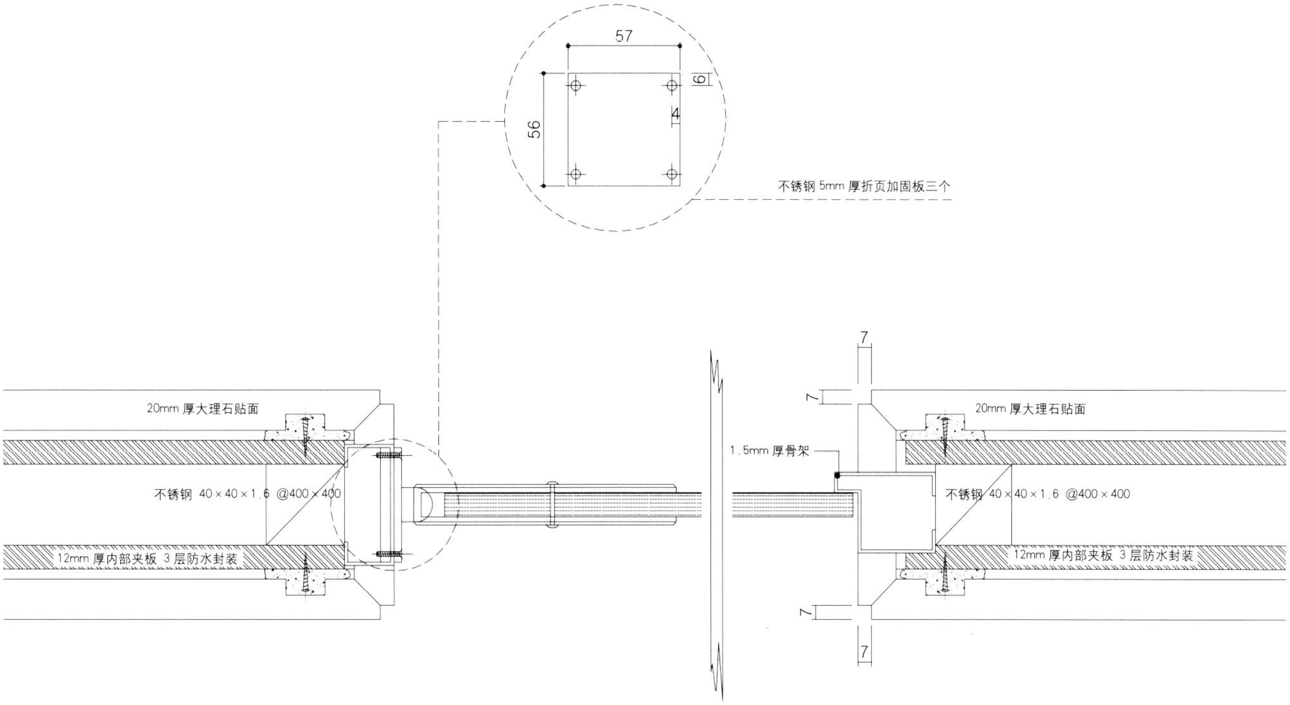

分隔门—剖面细部图（单位：mm）

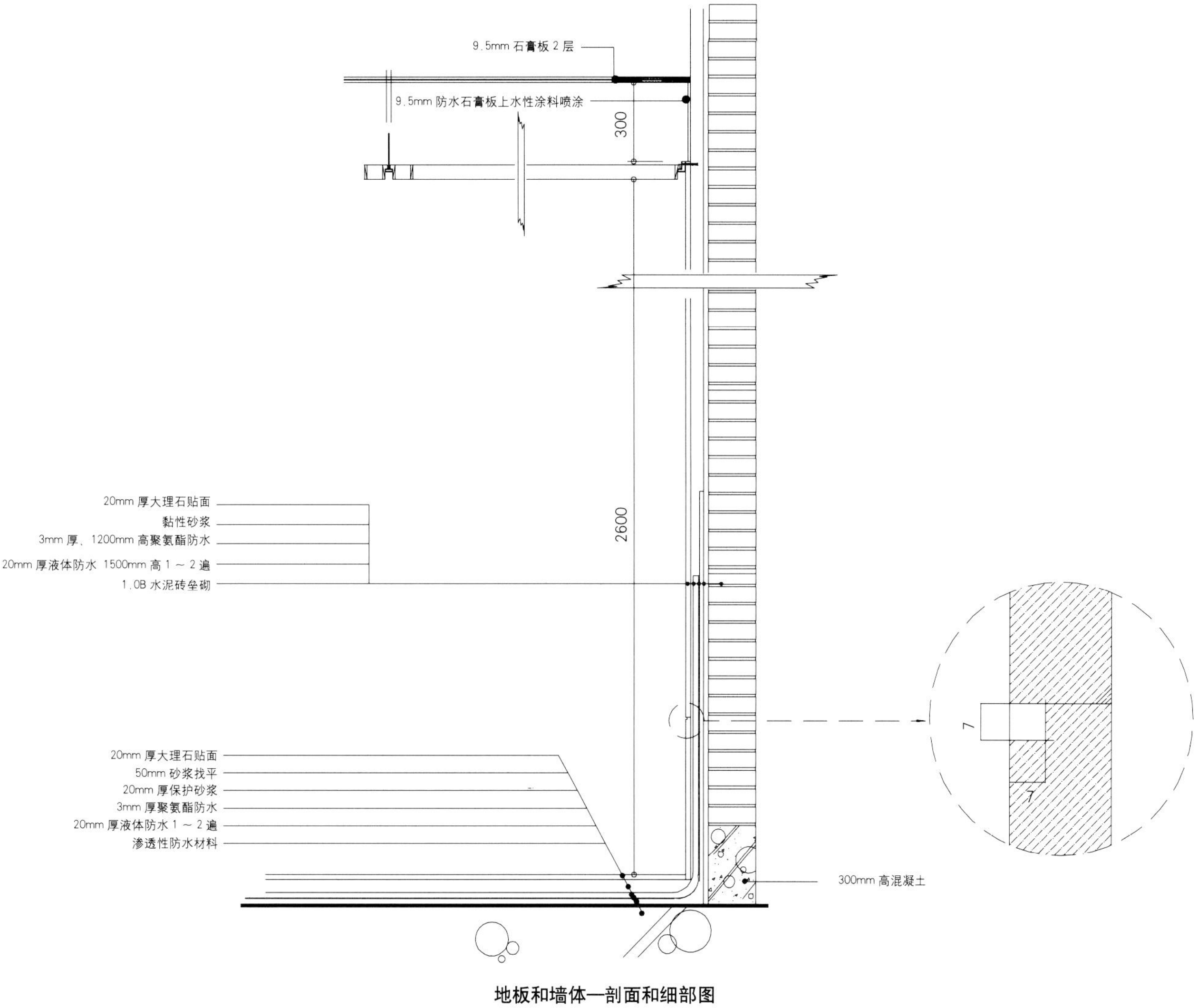

地板和墙体—剖面和细部图

Doriwon
桃梨园

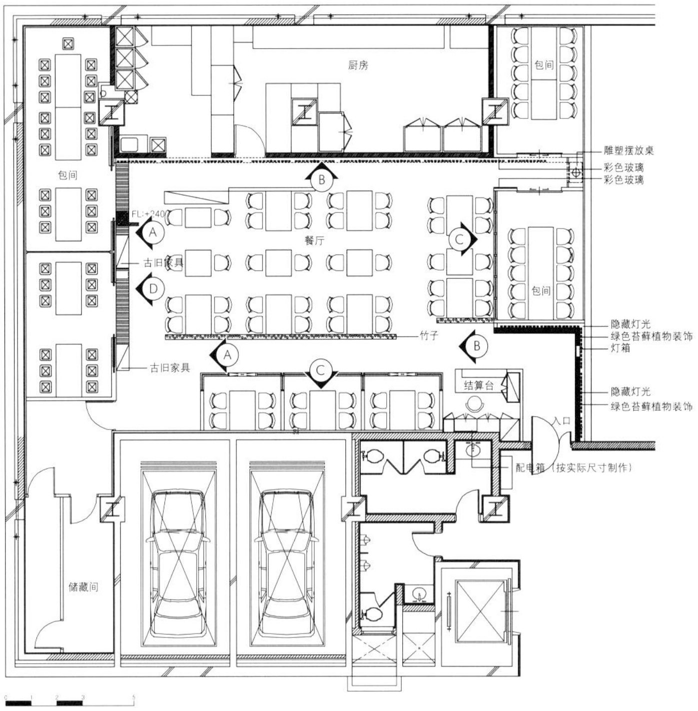

平面图

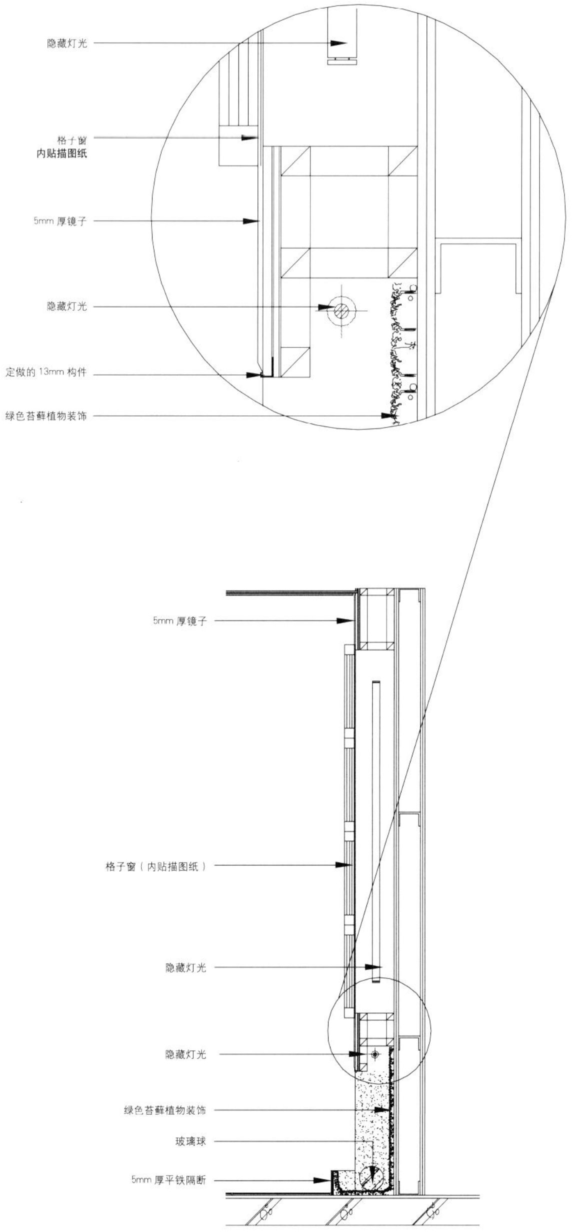

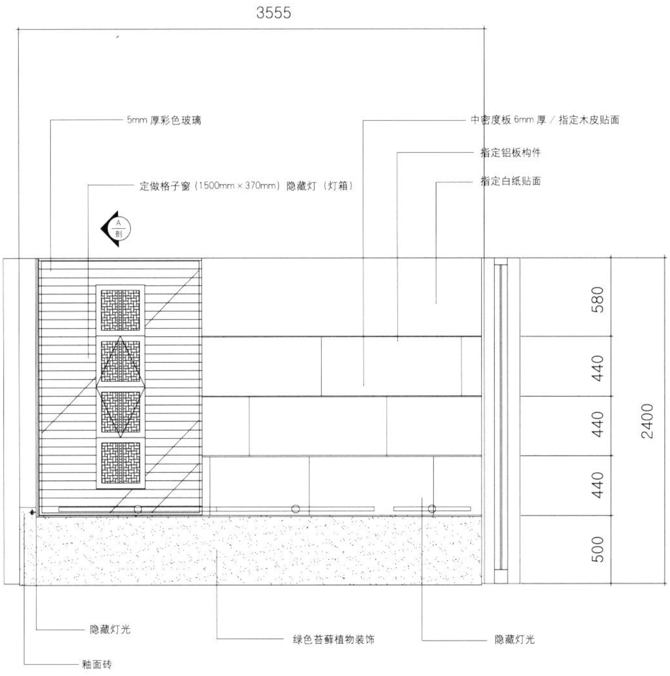

入口一立面图

剖面图 A

设计： A&A 设计公司 (02) 512—0121 | **施工：** A&A 设计公司 | **地址：** 江南区三成洞 | **面积：** 247.89m² | **装修：** 地面／实木地板 墙壁／涂料喷涂 顶棚／涂料喷涂

Design: A&A | Construction: A&A | **Built Area:** 247.89m² | **Finish:** Floor / Polishing Tile Wall / Painting Ceiling / Painting

■ 厅

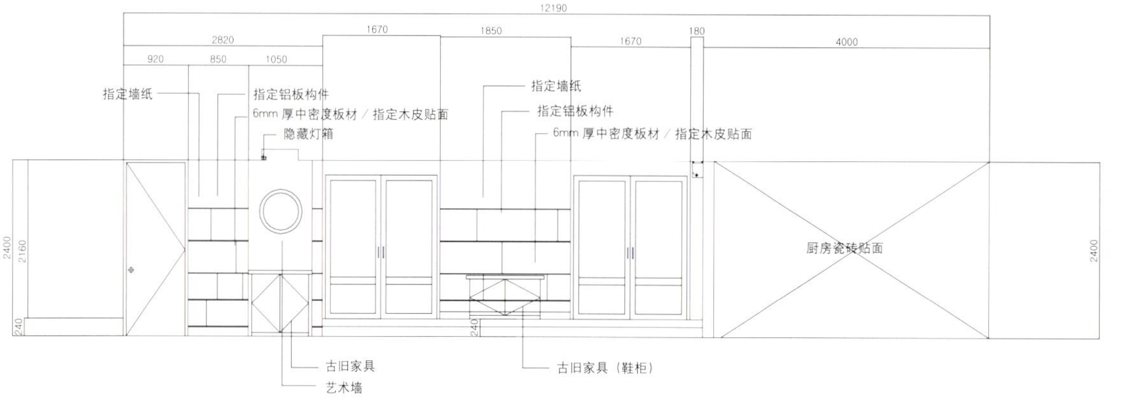

12190
2820
1670
1850
1670
180
4000
920
850
1050
指定墙纸
指定铝板构件
6mm 厚中密度板材／指定木皮贴面
隐藏灯箱
指定墙纸
指定铝板构件
6mm 厚中密度板材／指定木皮贴面
厨房瓷砖贴面
2400
2160
240
240
2400
古旧家具
艺术墙
古旧家具（鞋柜）

立面图 A

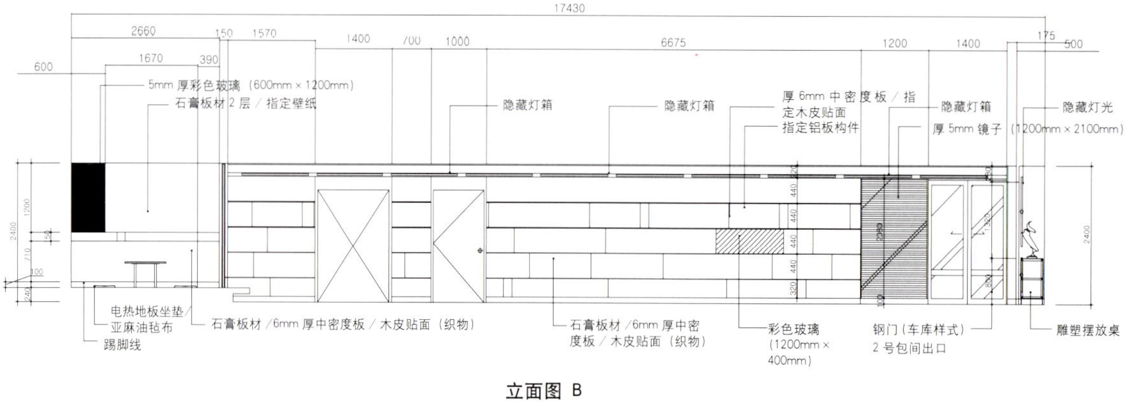

17430
2660
150
1570
1400
700
1000
6675
1200
1400
175
500
600
1670
390
5mm 厚彩色玻璃（600mm × 1200mm）
石膏板材 2 层／指定壁纸
隐藏灯箱
隐藏灯箱
厚 6mm 中密度板／指定木皮贴面
指定铝板构件
隐藏灯箱
厚 5mm 镜子（1200mm × 2100mm）
隐藏灯光
2400
电热地板坐垫／亚麻油毡布
踢脚线
石膏板材／6mm 厚中密度板／木皮贴面（织物）
石膏板材／6mm 厚中密度板／木皮贴面（织物）
彩色玻璃（1200mm × 400mm）
钢门（车库样式）
2 号包间出口
雕塑摆放桌

立面图 B

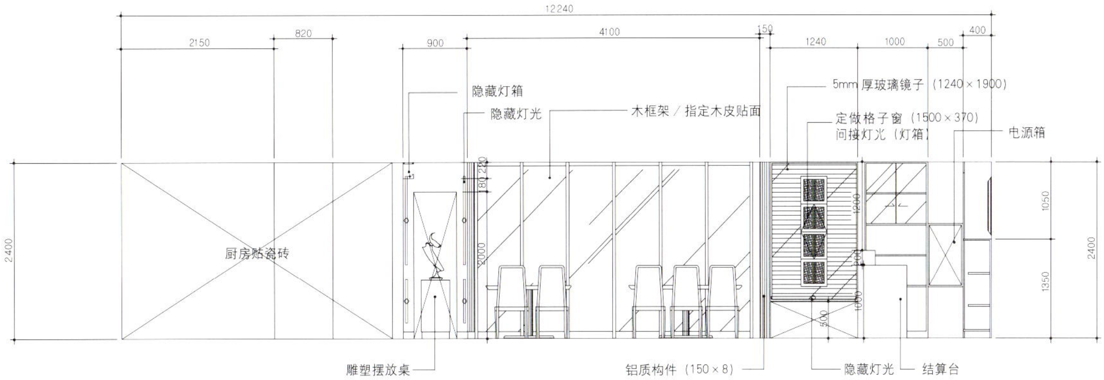

立面图 C（单位：mm）

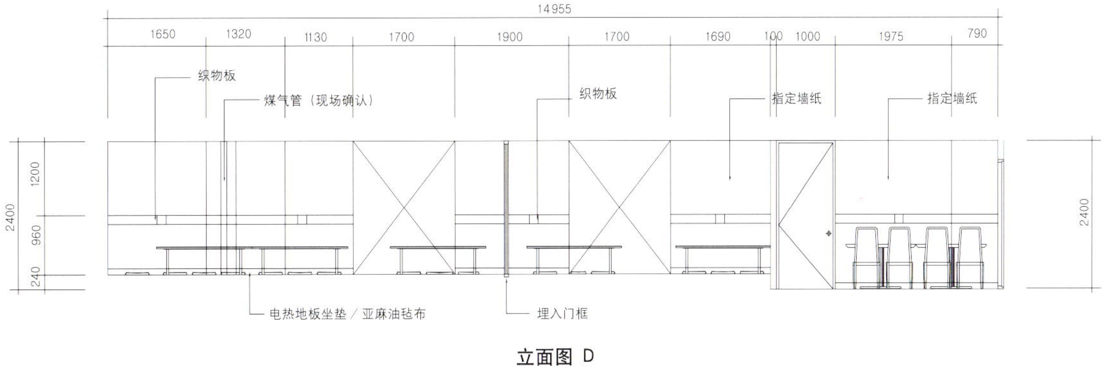

立面图 D

■ 单间

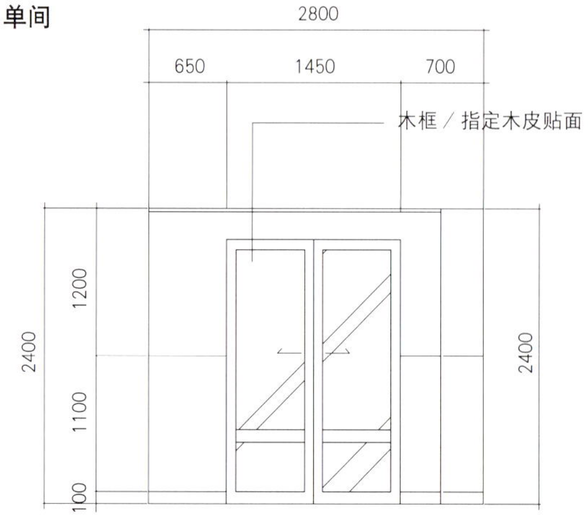

立面图 A

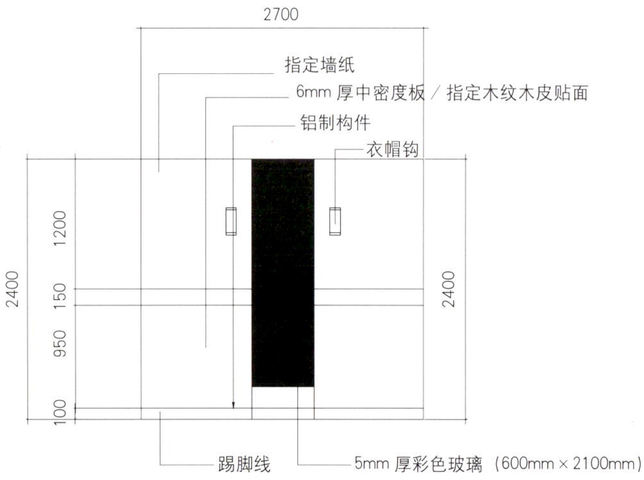

立面图 B

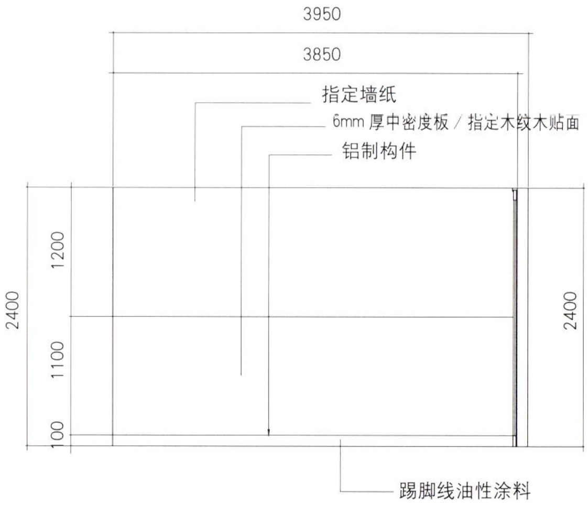

立面图 C

竹林幽草见边生，

晨森露珠挂叶枝。

最近，在康乐（Well–Being）时尚之风的影响下，我们以绿色有机农产品（译注：无公害蔬菜）原料为主，创建了健康饮食餐厅。

该餐厅有两个 10 人包厢和两个 4 人包厢，室内屏风采用了竖线条的玻璃隔断。沿着走廊的主要人流轴线两侧也采用了竹子的竖线条排列。走廊顶棚的反射光加强了走廊的悠长感，走廊的另一端放置了一件小型物件，从而形成一处小型展厅。其他的墙面与此相对比用横线条进行了强调，以 450 ~ 600mm 的间距放置铝板构件。地面、墙壁、顶棚都采用了白色涂装，装饰线条仅有垂直和水平两种方式，空间非常简约。

为了避免由于色彩单一造成的平庸感，在玻璃隔断房间的一侧墙面上通过竹绿色和樱桃红色的对比强调，活跃了房间的气氛，多盏球形玻璃吊灯也增加了房间的韵律感。

Fresh early morning of a bamboo field,

wearing crystal clear dew drops...

It is a fusion restaurant serving an organic agricultural products and neat atmosphere in a recent wave of 'well-being'.

Glass partitions of verticality divides two rooms for ten people and another two for four, and the long hallway also shows a repetition of vertical bamboo trees. The hallway has an indirect lighting box on its ceiling to emphasize its length, and at the end of it is a minimal objet, presenting a gallery-like space. We put slender aluminum moulding on the rest of the walls every 450-600mm, stressing the horizontal line. It is a simple space with outstanding vertical, horizontal lines on the floor, walls and ceiling.

We avoided boredom of the achromatic color, adding vitality by cherry-red wall of a glass-partitioned room, contrasted with green color, and the rhythm by glass balls of pendent lightings.

■ 走廊

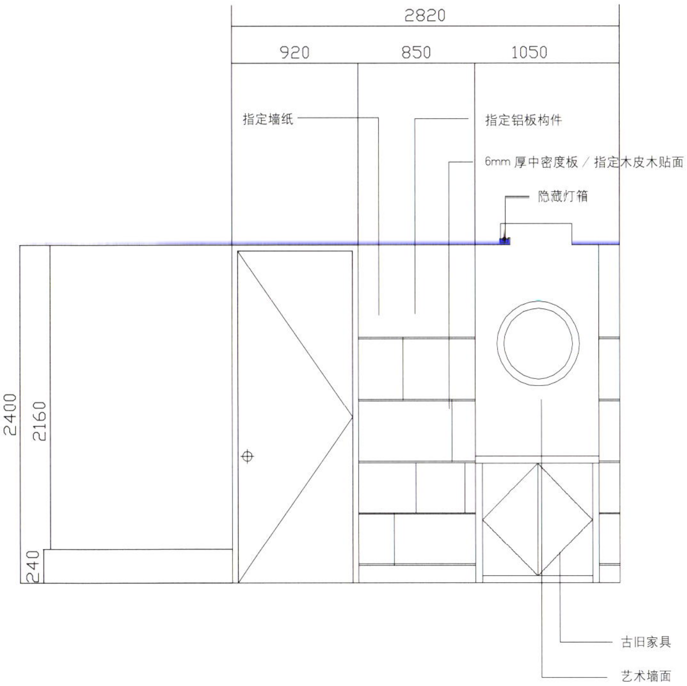

立面图 A

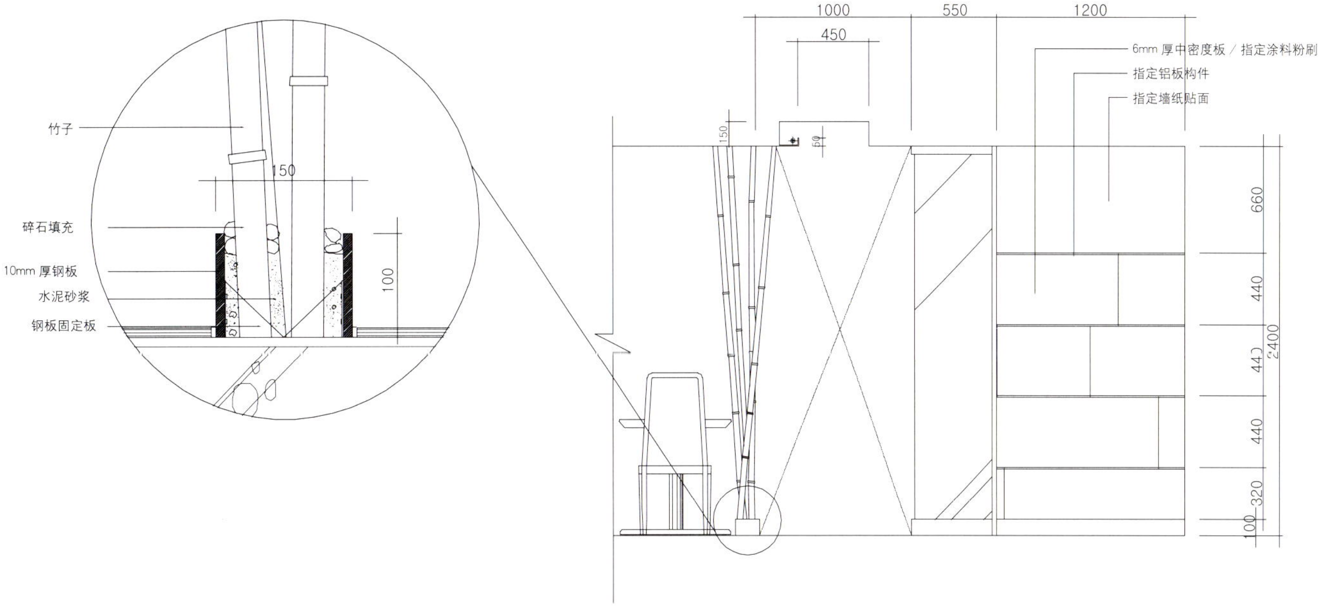

立面图 B 和细部图

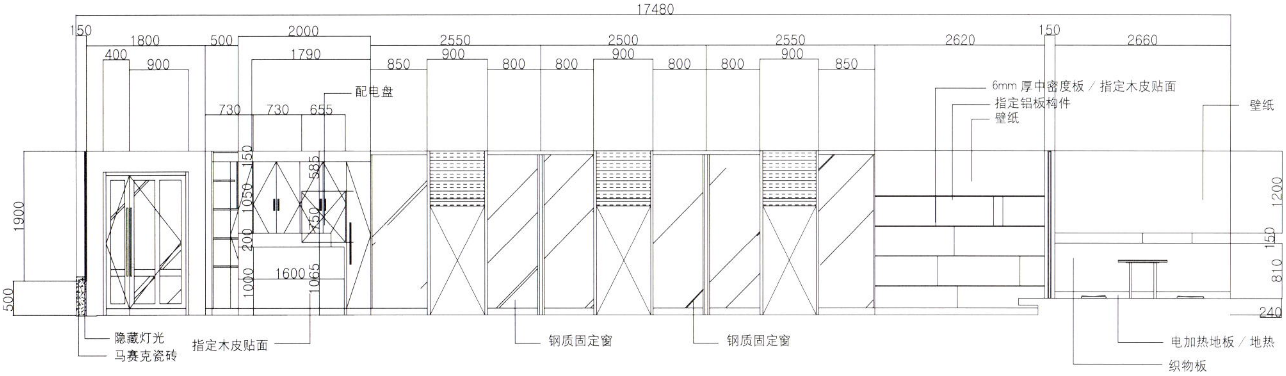

立面图 C

L'ABRI

辣不理餐厅

努力工作的您，旅游去吧！这是夏天假期到来时，常常听到的劝告。就像用蓝色遮住了熊熊燃烧的太阳光，即使稍微想像一下也会觉得忽然间凉爽了许多。20 世纪 70 年代位于韩国江陵市镜浦台的高级商务宾馆很景气，但是，随着岁月的流逝，这种情况悄悄发生了变化，由于客源不足等原因，不得不进行更新改造或扩建。当然，再装修是按照与以往不同的方式，以更加注重时尚的方式进行的。

与酒店大厅横竖简单呆板的划分不同。该餐厅采用了大尺度的曲线构件将室内空间扭曲变形，使人们产生了空间变化的错觉，从而提升了空间的品位。室内光照也像变色龙一样，室内灯光与窗外的自然光线结合在一起，变化演绎出多彩的室内效果，白天和黑夜各不相同，白天呈现黄色，夜晚呈现蓝色。

室内地面用地毯和地砖进行了分区铺设，空间的归属感更强，将立面上横竖的呆板生硬化解了。

位于入口处侧面的大型落地窗与自然环境融为一体，富于活力。真凉快！自然赋予的丰富感与落地窗吸纳的外部环境要素以及太阳光等一次性地映入眼帘，使人感到非常惬意。红色装饰玻璃给予空间以活力，屏风与木质桌面也用现代的设计手法进行了处理。同时，五种不同风格的椅子与自然环境协调统一更加丰富了室内空间。真惬意！这里是追求快节奏的现代都市人的小憩空间。人们能在此体会到没有压力、没有紧张、没有烦躁的自由自在。设计师希望通过这一独特的设计，使客人静思冥想，从而忘记工作生活的烦恼，体会一下完全放松的感觉。

'Take a trip, hard-worker!' These words come up more closely as the summer vacation season nears. A blazing sun and waves that would cover up everything with blue. Coolness is strongly felt just thinking of it.

The hotel, which had been the best business hotel in Gyeongpodae until 70s, eventually went under large-scale renovation, enlargement and extension. The interior focuses on differentiating the design from the existing one.

Amusement is found in the contrast between the lobby of rigid vertical, horizontal line and the curved spatial structure. It is like a chameleon. The space changes its looks by an artificial illumination and natural light coming in through the window. Yellow during the day and blue at night. It would be quite fun to watch it changing its clothes.

As for the floor, a carpet and flooring divides the space in a natural way, thus softening the vertical, horizontal space that can easily be stiff.

Coolness. From the entrance, a large side window joins nature and offer vitality. Along with spaciousness, the daylight and natural elements pour down from outside through a large window. A glass wall of decorative red adds life to the space, and a formative element, which comes from a wood pattern on partitions and tables, is expressed in modern style. And five different sofas bring harmony with natural color, making the space rich. Leeway. This place will provide relaxation to modern people who stresses on quickness. I hope they fall in love with a unique attraction of this place, fully enjoying their time.

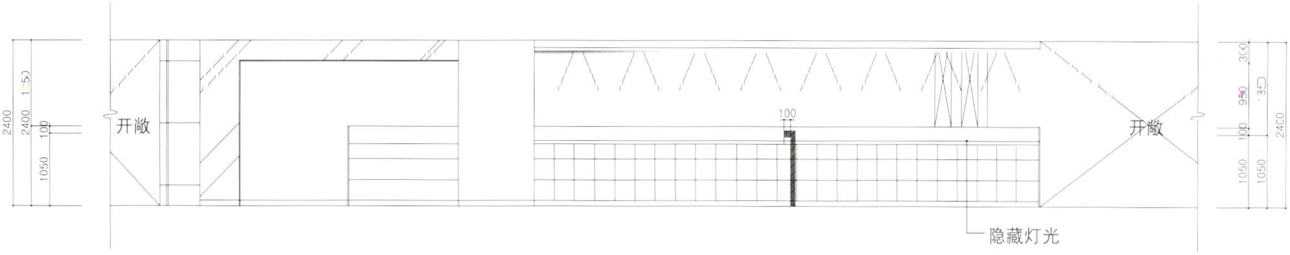

厅一立面图 A

设计： PDA 设计公司 黄成勋、崔英玉、刘秀然、林英爱、孙荣恩、金弦真、宋恩美（02）792-4206 | **施工：** PDA 设计公司 于然、张哲、梁正植、朴闽州、林正亚、林韩亚、林海申、申商度、黎则勇、姜梁丘、申尚玄 | **地址：** 江原道江陵市江门洞 303-4 | **面积：** 1181m² | **装修：** 地面 / 大理石、地毯 墙面 / 大理石、木纹木、壁纸 顶棚 / 聚酯喷涂、壁纸 | **摄影：** 金韩秀

Design: PDA (Phillip, David & Associate), Phillip · Hwang, Sung hoon / Corinne Chio / Yoo, Soo youn / Lim, Young ae / Son, Young eun / Kim, Hyun jin / Song, Eun mi | **Construction:** PDA · Woo, Youn / Jang, Chul / Yang, Jung sik / Park, Min joo / Lim, Jung ah / Lim, Han na / Lim, Hea sine / Shin, Sang du / J.Y.Lee / Kang, Ryang goo / Sin, Sang hyun | **Built Area:** 1181m² | **Finish:** Floor / Marble, Carpet Wall / Marble, Veneer, Wallpapering Ceiling / V.P, Wallpapering | **Photograph:** Kim, Han soo

L'ABRI

L'ABRI

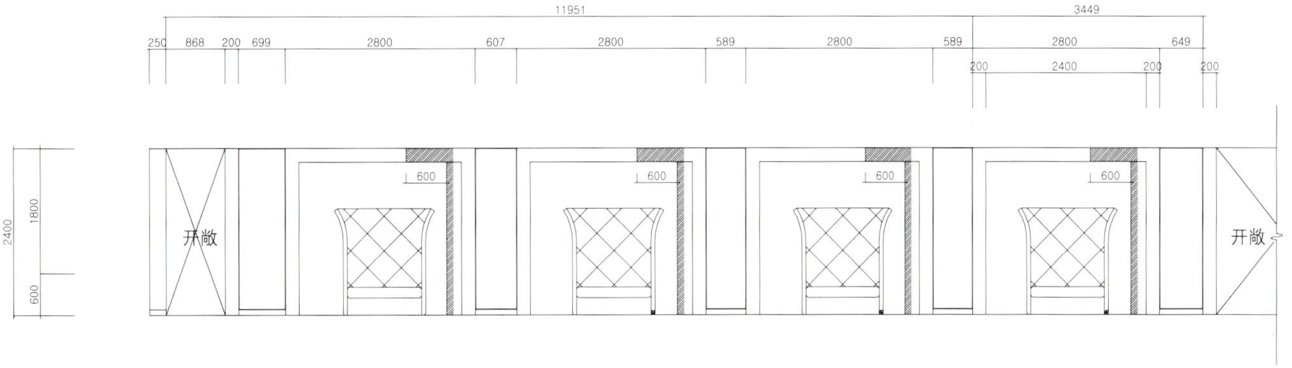
11951
3449
250
868
200
699
2800
607
2800
589
2800
589
2800
649
200
2400
200
200
600
600
600
600
开敞
开敞
2400
1800
600

厅一立面图 B

L'ABRI

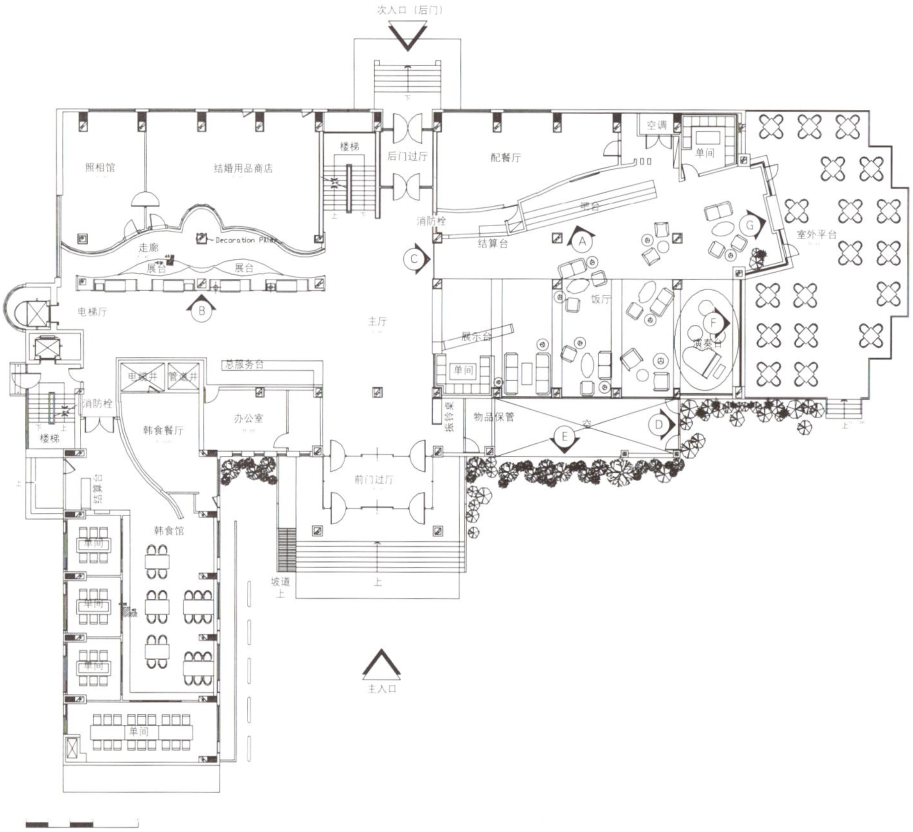
次入口（后门）
楼梯
后门过厅
配餐厅
空调
单间
照相馆
结婚用品商店
吧台
消防栓
结算台
室外平台
走廊
展台
展台
Decoration Panel
电梯厅
主厅
饭厅
展示台
单间
演奏台
总服务台
电梯井
管道井
消防栓
办公室
韩食餐厅
楼梯
物品保管
前门过厅
韩食馆
单间
单间
单间
单间
坡道
上
主入口

平面图

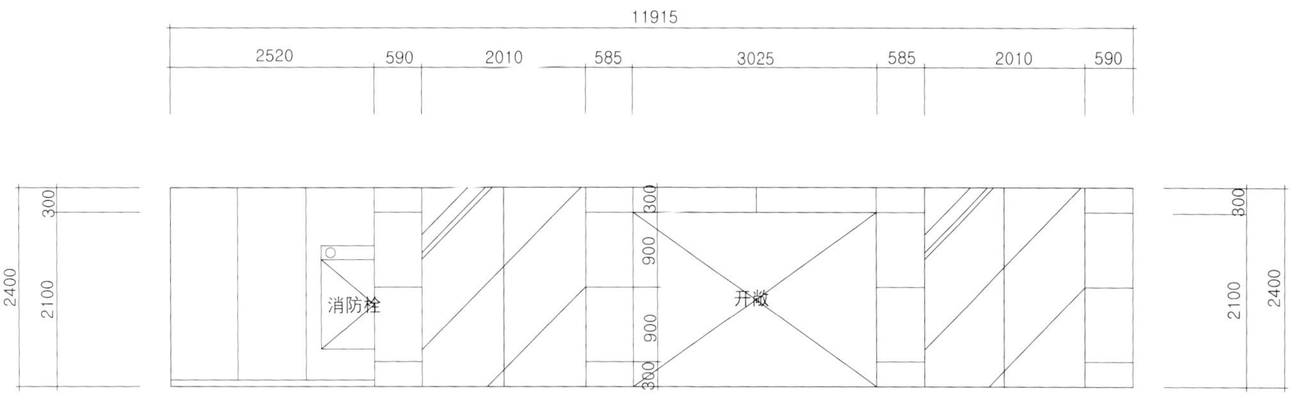

厅一立面图 C

■ 饭厅

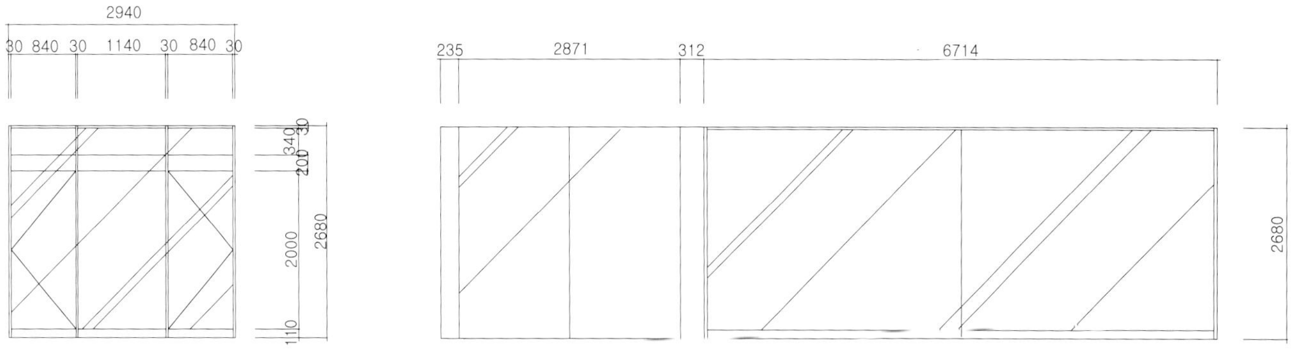

立面图 D、E

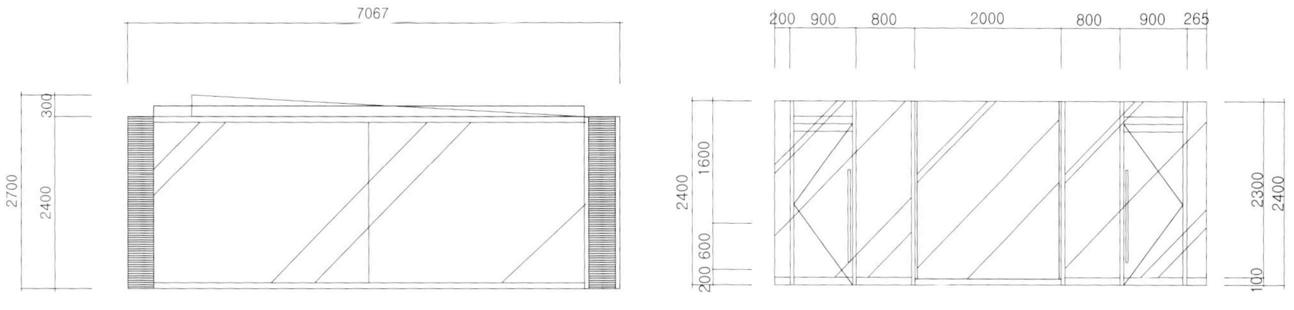

立面图 F、G

■ 饭厅—造型墙

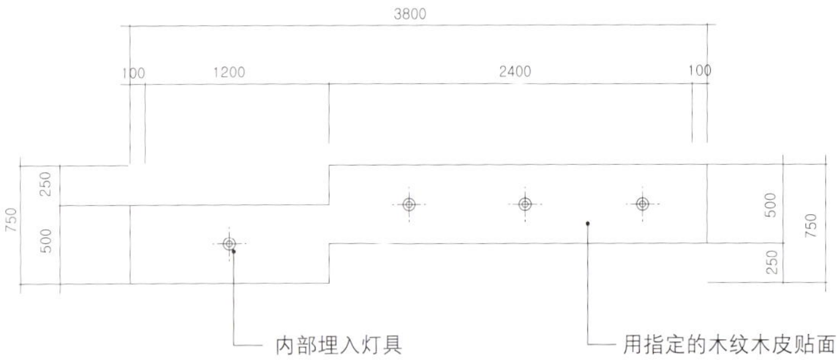

顶视图（平面图）

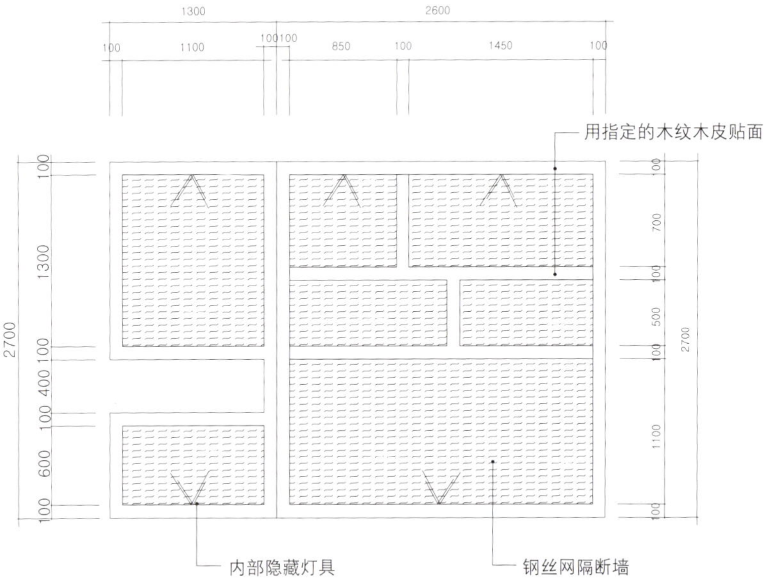

前视图（正立面图）

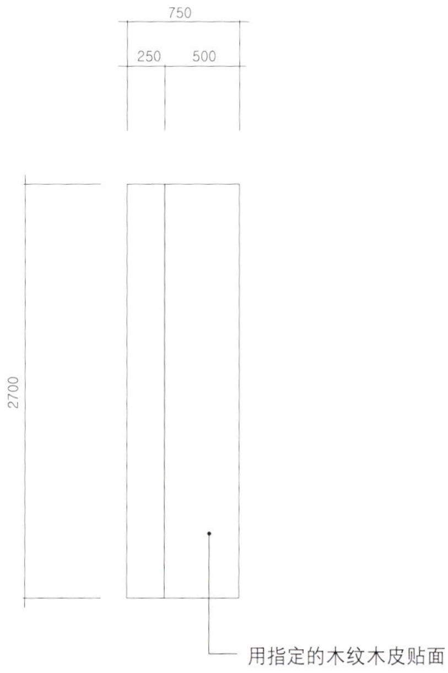

侧视图（侧立面图）

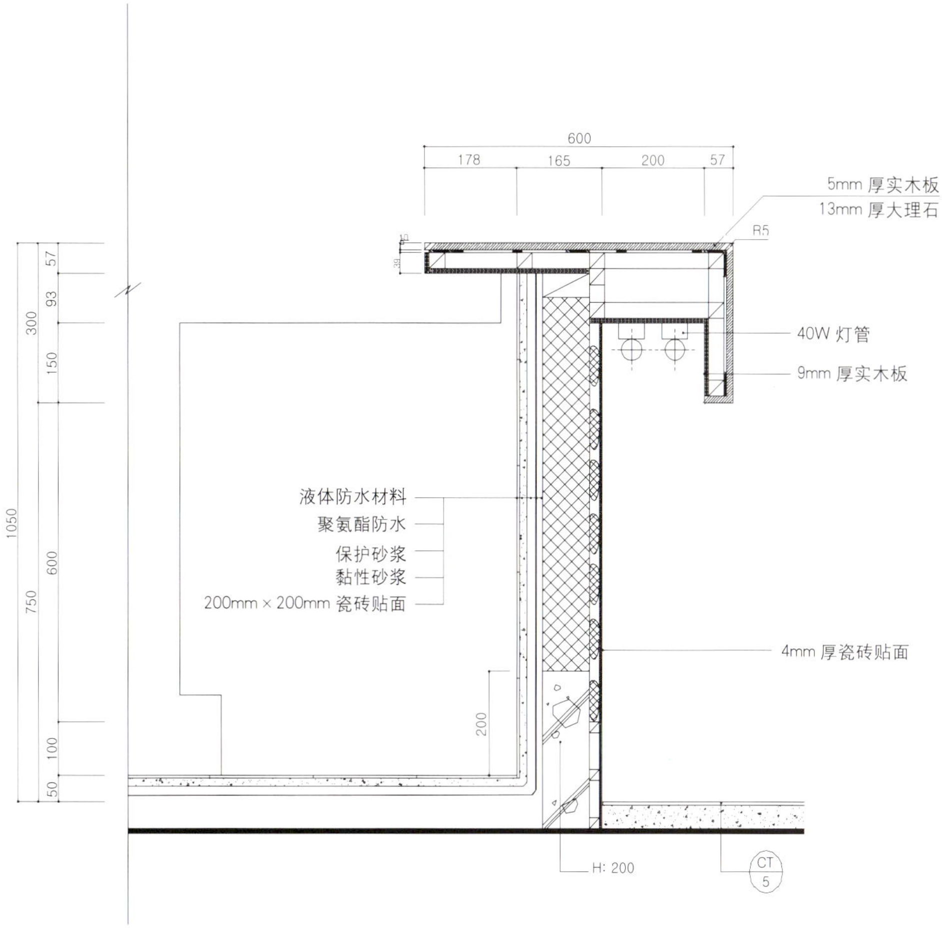
600
178
165
200
57
5mm 厚实木板
13mm 厚大理石
R5
40W 灯管
9mm 厚实木板
液体防水材料
聚氨酯防水
保护砂浆
黏性砂浆
200mm × 200mm 瓷砖贴面
4mm 厚瓷砖贴面
1050
300
750
57
93
150
600
100
50
200
H: 200
CT
5

结算台侧剖面图

Tomo

陶茂饭店

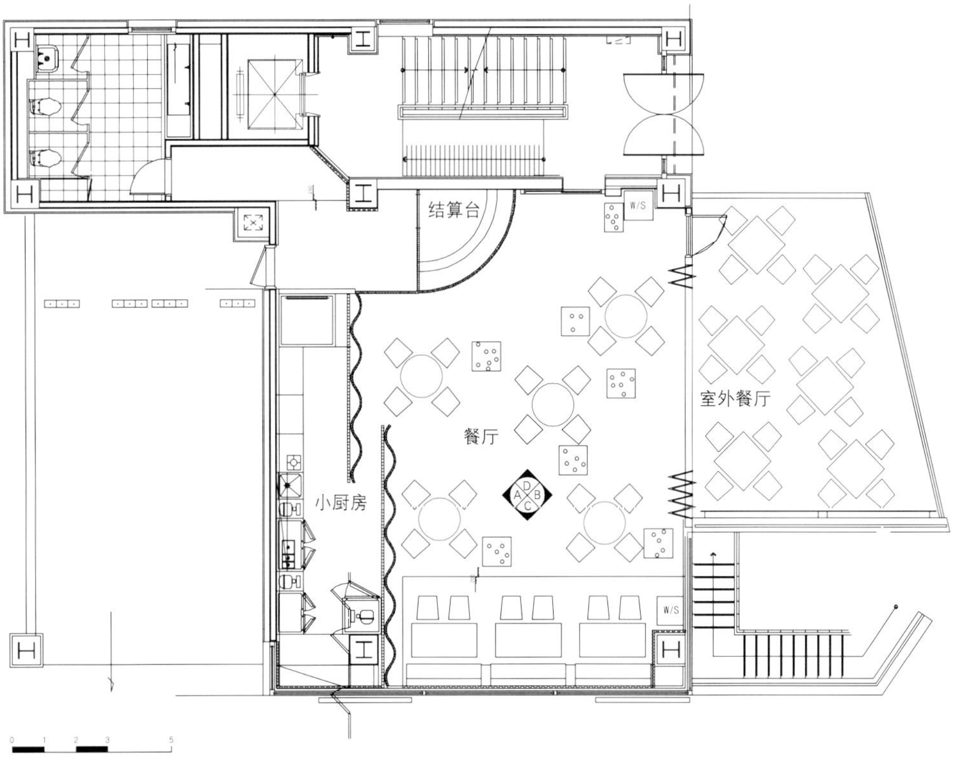

一层平面图

华丽的霓虹灯和聒噪的音乐声，满大街的摇滚酒吧，这是弘益大学附近的一条特色步行街。街道一侧阳台的墙面上挂着巨幅写实竹子装饰画，这使得漫步的行人驻足观望，脑中似乎掠过“真想稍憩片刻，感受一下竹林里的凉风”的想法。这种想法被设计师原封不动移到了室内。如果说在外部能够感受到竹林的话，一层内部的竹灯照明也使人有这种温暖的感觉。这种竹灯照明能否使人感受到“健康”呢？设计师非常小心地说起过。不用顶棚的照明，而是采用室内模拟竹子的照明手法，激发了客人的好奇心。用珍珠幕帘将职员空间隔离出去，它不仅遮挡了视线，也是一种装饰手法，能消除狭小空间的沉闷感，显然比用隔墙优越。如果说一层是利用外部光线形成了开放空间，二层则是通过似断非断的编织物形成了半封闭空间。竹子榻榻米和编织物的装饰构成了日本式酒馆的特点，设计师曾经坐在榻榻米房间内注视着红色的墙壁，对此狭窄悠长的空间应该赋予何种主题很伤了脑筋，最后，还是决定为情侣们而设计。对于设计师和雇主来讲弘益大学陶茂饭店这么安排等于是冒险。虽然，鸭欧亭店和清潭店（这两个饭店都位于首尔的富人区。——译者注）都很成功，但是，位于弘益大学附近的这处“陶茂”能否让顾客接受却没有把握。低矮的层高、相互分离的上下层以及糟乱的外部环境都是摆在金光泰所长面前要考虑的问题。金所长经过多次尝试而发明的竹子照明现在已经成了陶茂饭店的标志，从他微笑的言谈中也可以感觉到轻松和愉快。

Flamboyant neon signs and loud music... An alley near Hongik University lined with rock bars... People stop to look at the real print of bamboo forest picture decorating a side of the terrace, and think ‘It must be a great to place to drop by and feel the cool wind of bamboo forest’. The cool atmosphere of the terrace is transferred intactly to the interior, while the first floor bamboo illumination emits a warm feeling. The designer defines it is a feel of ‘fusion’ that comes from the lighting pillars installed hear and there. To which, the ceiling illumination gives way to guide people to curiosity. The almost transparent bead screenings remove the closedness from a small space, where a space for staff is hidden beyond a chain screening, which functions as a decoration and not just as a veil. The first floor is an open space, and on the other hand, the second floor is a closed one. Bamboo tatami and a fabric screening suggest a Japanese pub. A red wall of a tatami room... The designer pondered on the function of this long, narrow space, and decided to present an amusing space for couples. Tomo, Hongdae may have been an adventure both to the designer and client. Apgujeong and Cheongdam branch were success, but it was not sure if it would appeal to people in a college district like Hongdae. A low ceiling, separated floors and the surroundings... Director Kim Gwang had to deal with all these tasks. The bamboo lighting constructed after multiple failures, he says with a smile, has become a representative image of Tomo.

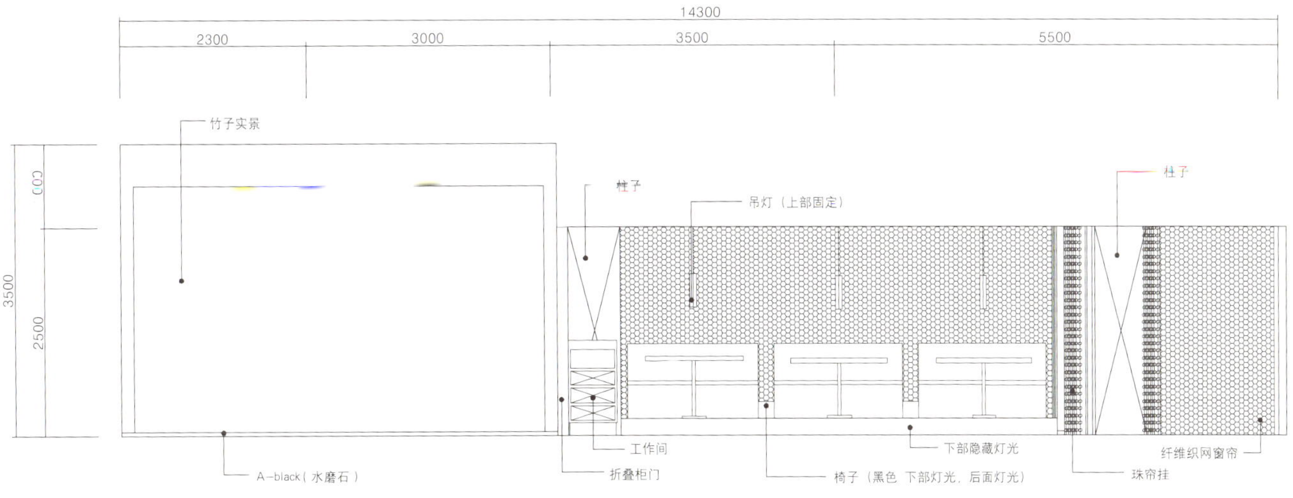

一层餐厅一立面图 A

设计：（株）太和综合建筑师事务所 金光泰、姜成文、安镇城、姜真浩 (02) 518-0080 | **施工：**（株）太和综合建筑师事务所 姜成文、林才丛 | **地址：**首尔市麻浦区西桥洞 409-1 番地 | **面积：**一层：139m²，二层：206m² | **装修：**地面 /AIC 板、竹子、彩色玻璃、大理石 墙面 / 织物、金属网、球链、彩色漆喷涂 顶棚 / 外露、蛭石粉喷涂、粉刷

Design: Taeha Kim Architects Co., Ltd. · Kim, Kwang tae / Kang Sung mun / An, Jin sung / kang, Jin ho | **Construction:** Taeha Kim Architects Co., Ltd. · Kang, Seong mun / Im, Chae chung | **Built Area:** 1F -139m², 2F - 206m² | **Finish:** Floor / ALC Panel, Bamboo, Color Glass, Marble Wall / Fabric, Metal Mesh, Ball Chain, Color Lacquer Ceiling / Exposed Ceiling, Vermiculite Spray, Painting

■ 一层餐厅

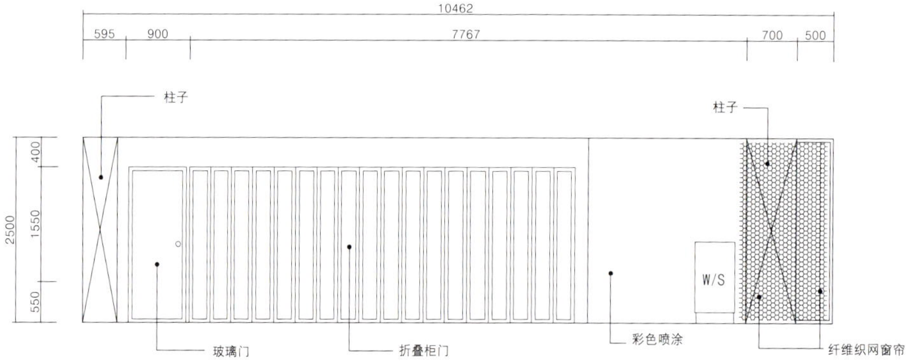

立面图 B

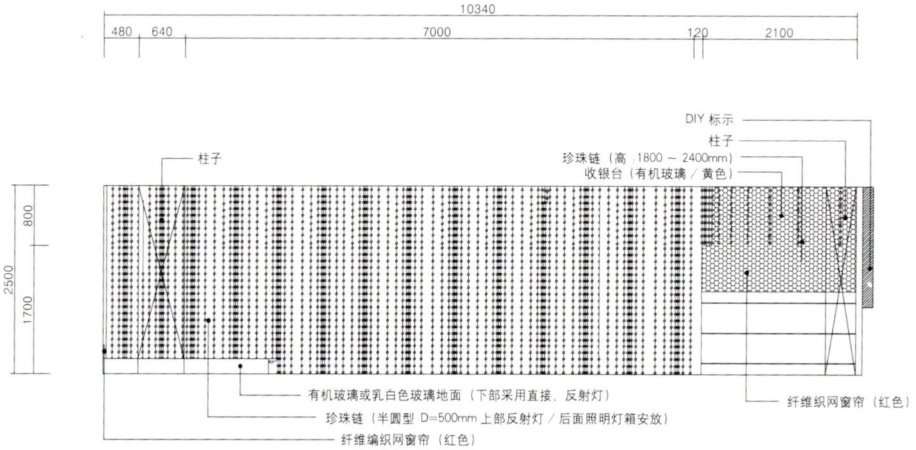

立面图 C

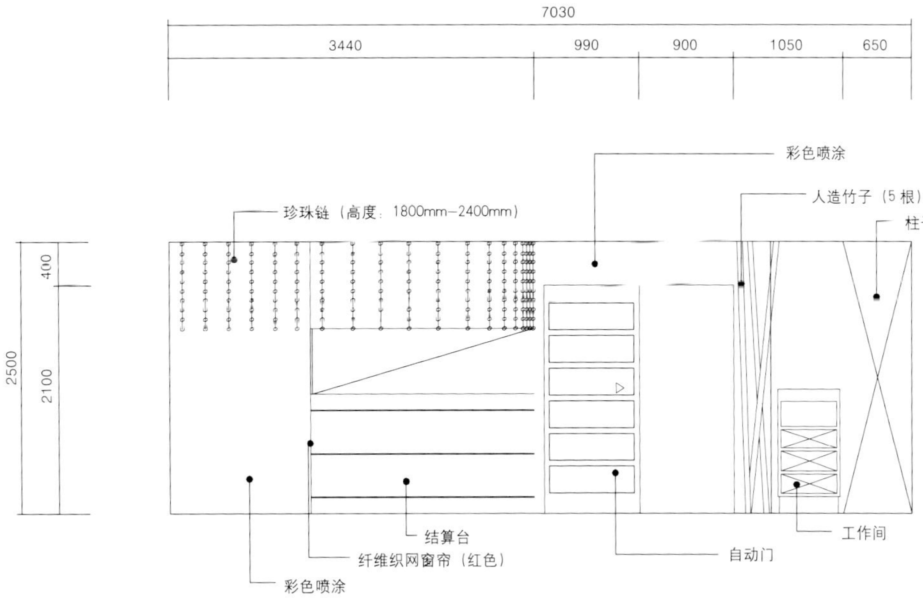

立面图 D

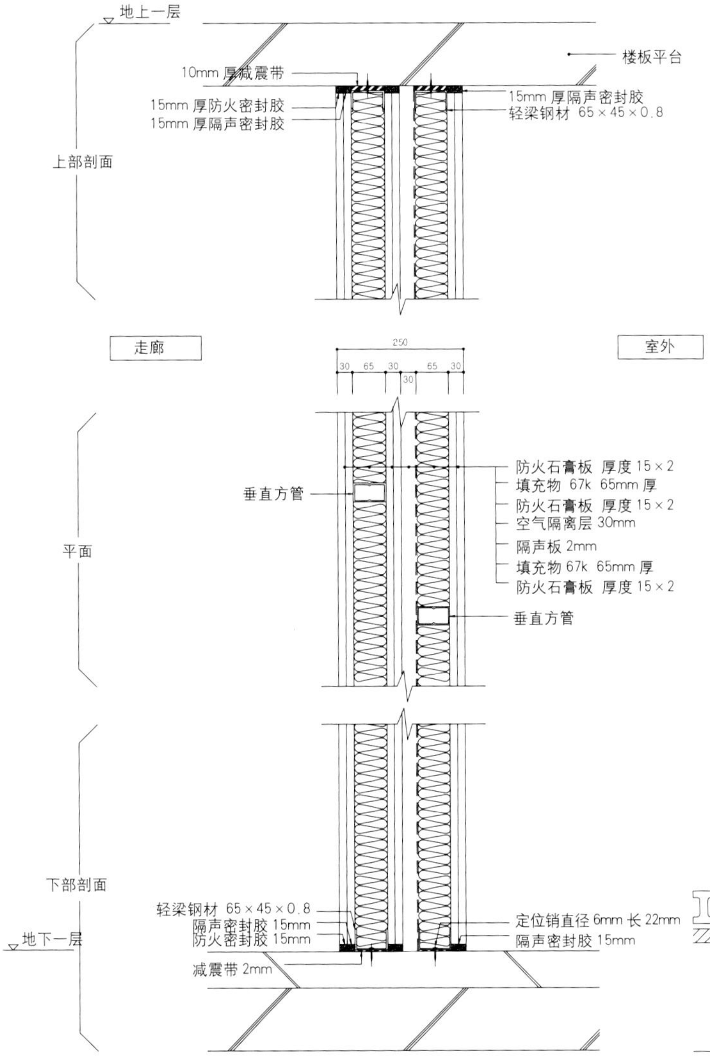

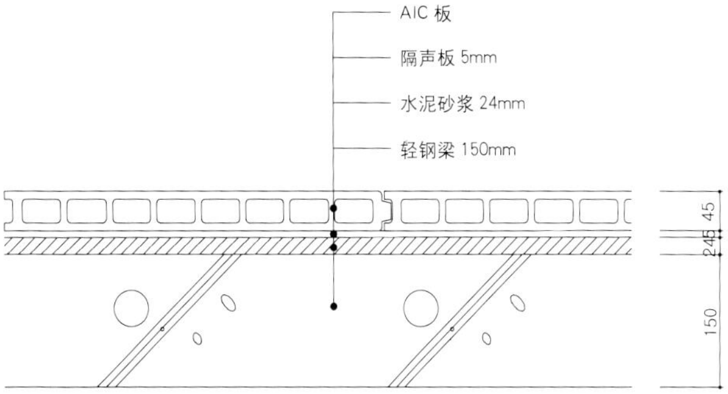

墙体一剖面图（单位：mm）

地板一剖面图

■ 二层餐厅

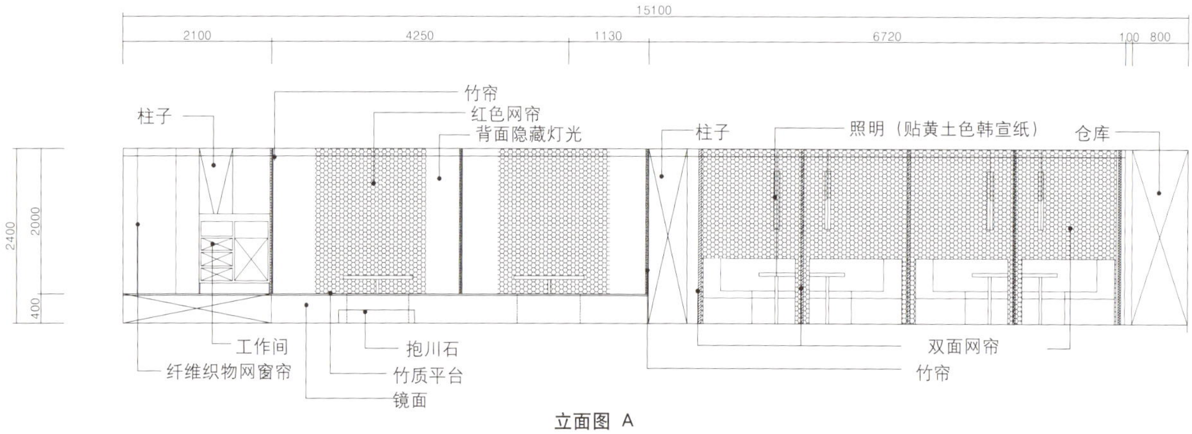

立面图 A

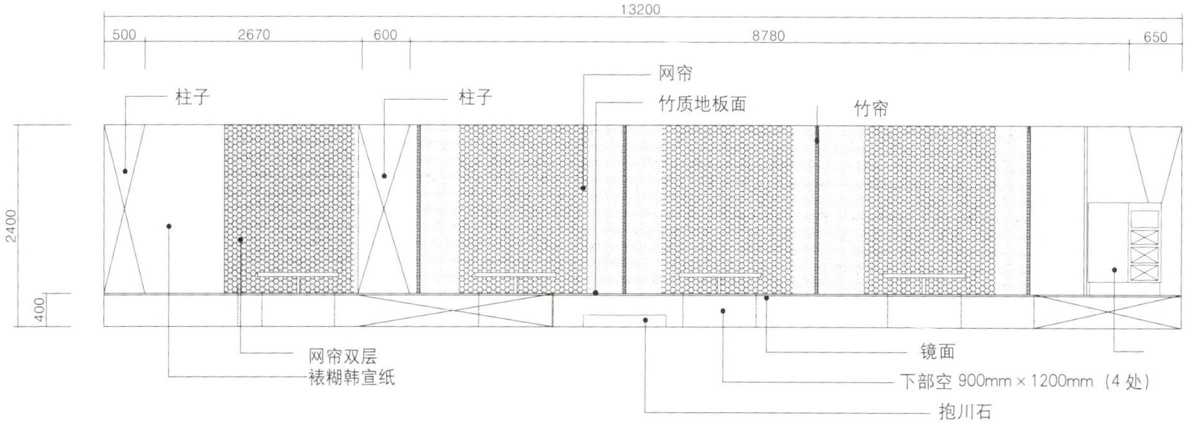

立面图 B

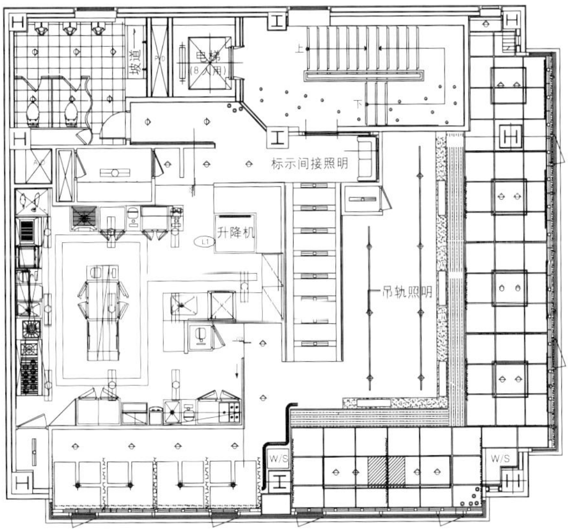

二层顶棚平面

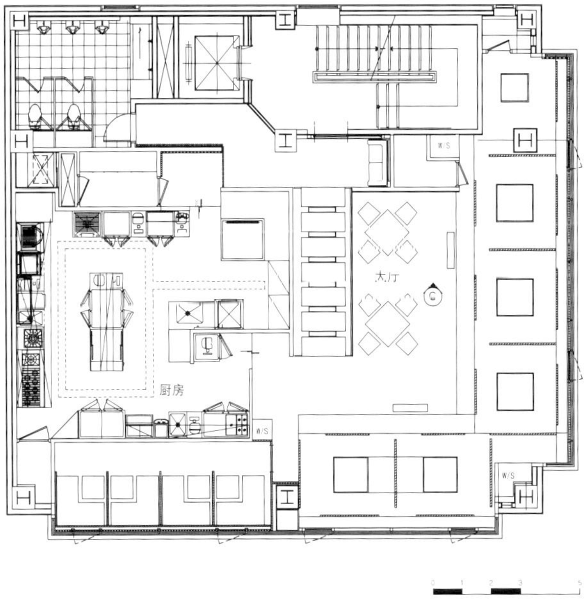

二层平面

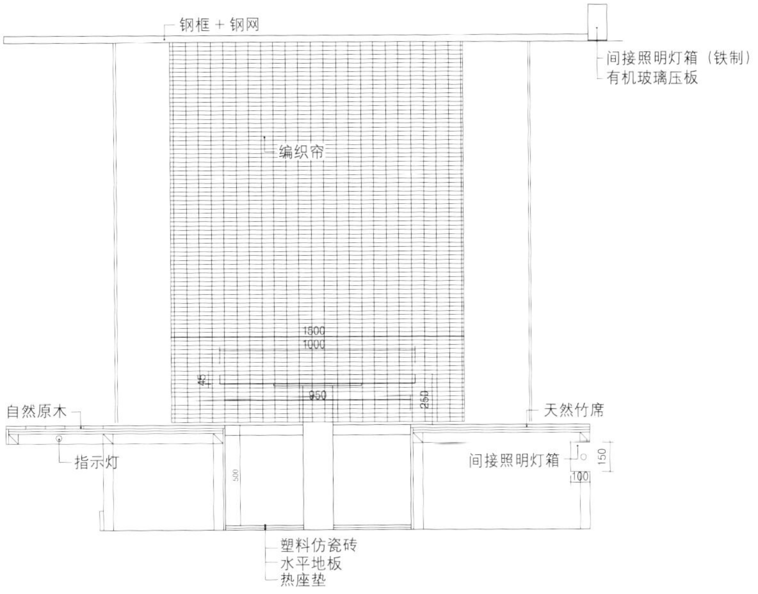

榻榻米—细部图

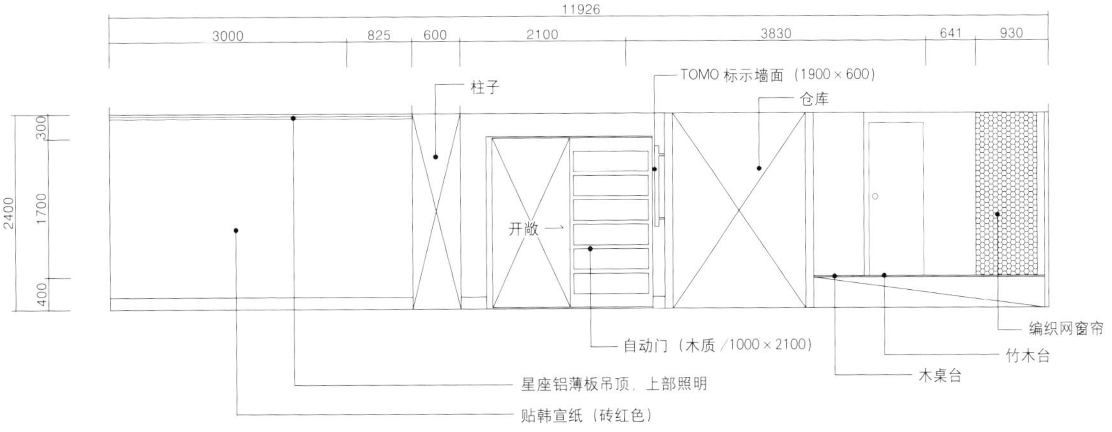

立面图 C（单位：mm）

Al Parco

阿巴洛克餐馆

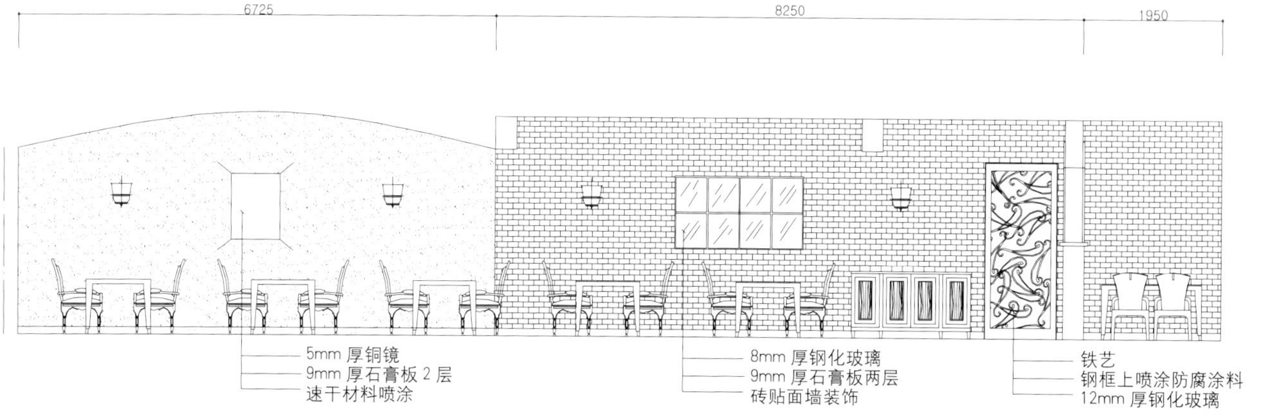

一层餐厅一立面图 A

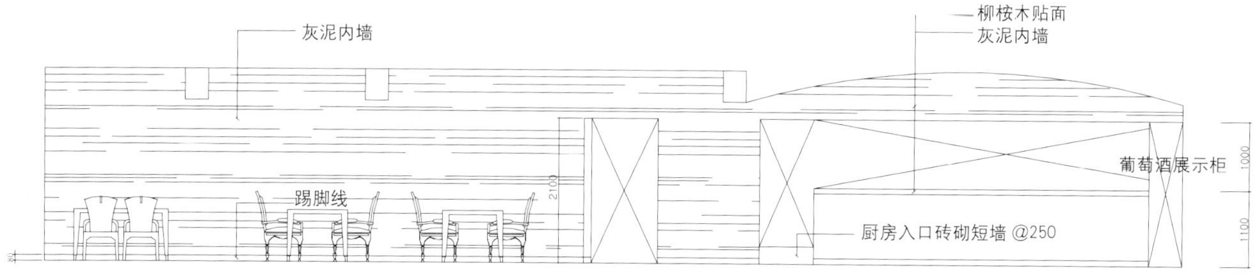

一层餐厅一立面图 B

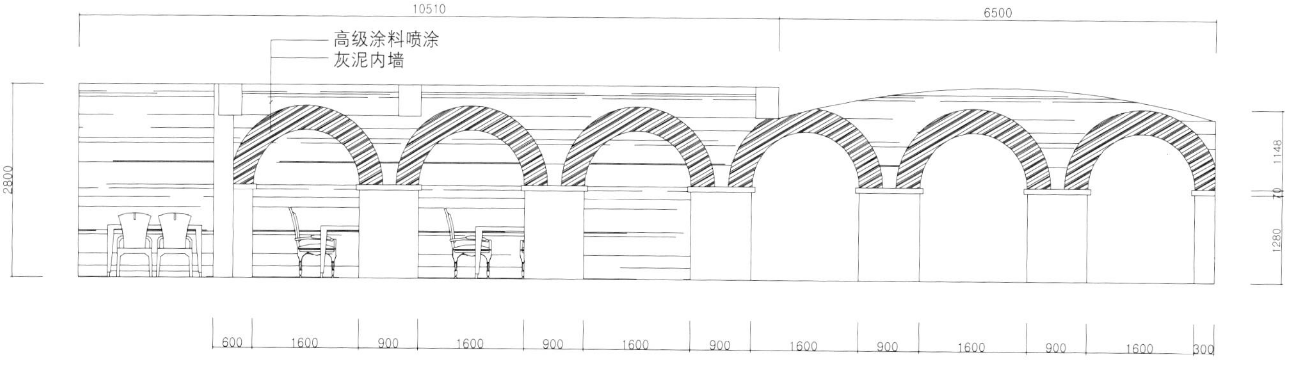

一层餐厅一立面图 C

一般地说，销售意大利面食和葡萄酒的传统饮食店，雇主对其特色要求甚高。建筑的正立面采用了旧砖块以表达岁月悠久的古朴之感，在选择商店标牌时，采用了松秋古董店里的旧材料并对每一个字母都一一进行了腐蚀加工，通过这些锈迹斑斑的字母，古朴的风格和异国的情调活灵活现地得到了展现。

地下一层借用了洞穴的概念，地面采用了鹅卵石铺地，室内尽量采用局部照明而没有采用集中的大灯照明。地面层通过与室外空间的渗透，创造了露天自助餐餐厅一样的效果，它的设计理念是豁然开朗的城市广场。大厅周边的列柱不仅划分了空间也散发了异国情调，同时，这一周圈的列柱也作为广场的一部分展现在客人面前。二层是意大利葡萄酒特色吧台，虽然，意大利葡萄酒的历史和传统很悠长，但是，韩国人了解较少，认知度比较低。因此，该酒吧特意强调了悠久性和特色，利用原木、编织物、石料等天然材料，将空间划分成一定的区域。

设计： KcmD 设计公司 郑石然、李秀允、蒋韩荣、边宝庆 (02) 3443—1071 | **施工：** KcmD 设计公司 郑石然、李凡虎 | **地址：** 江南区新沙洞 643—2 汇强大厦 | **面积：** 地下一层：171m^2、地上一层：206m^2、地上二层：156m^2 | **装修：** 地面 / 石材瓷砖、环氧树脂、雕刻 墙面 / 碎砖块、速干涂料、灰泥、仿石材涂料、防腐涂料、织物、花岗石、条石 顶棚 / 彩色喷涂、水性涂料 | **合作部门：** 木工 / 白庆万 五金 / 巨陆金属公司 涂墙 / 善明植 玻璃 / 萨罗特克莱斯公司 照明 / 国土照明公司 设备 / 大东设备

Design: KcmD Co., Ltd. · Jeong, Seok yeon / Lee,Su yun / Gang, Han yeong / Byeon,Bo gyeong l **Construction:** KcmD Co., Ltd. · Jeong, Seok yeon / Lee, Beom ho l **Built Area:** B1F 171m^2, 1F 206m^2, 2F 156m^2 l **Finish:** Floor / Stonewear Tile, Epoxy, Mother of Pearl Wall / Used Brick, Dry-vit, Antico Stucco, Stone Paint, Corrosion Paint, Fabric, Granite, Slate Stone Cedar Wood Ceiling / Color Lacquer, Water Lacquer

■ 正立面

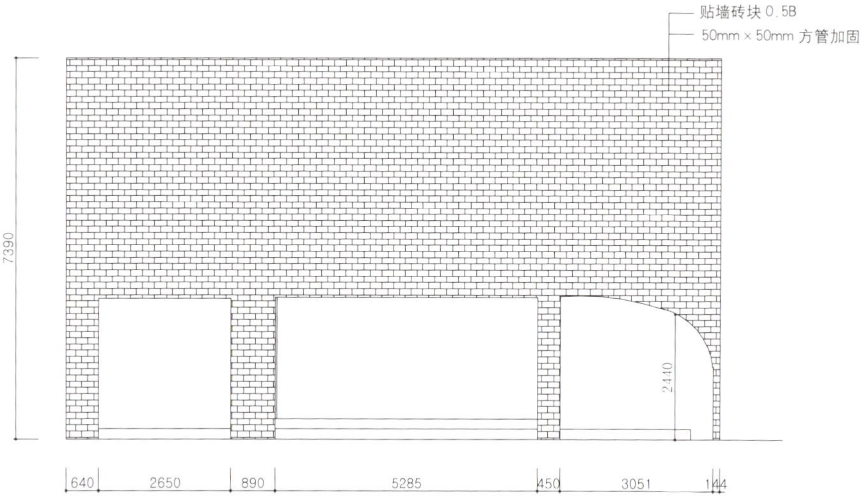

正立面图

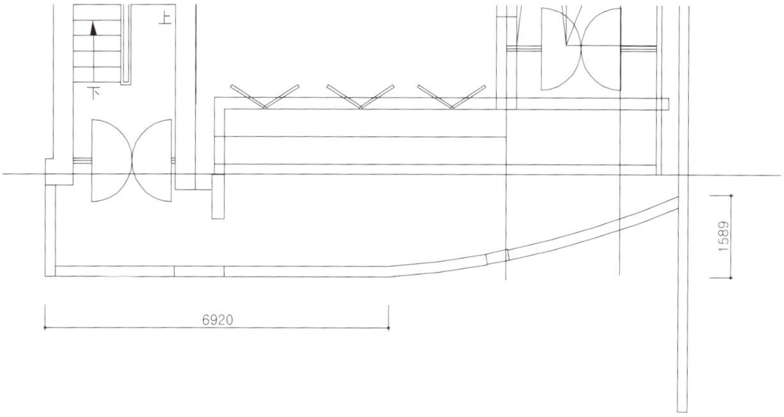

局部平面图

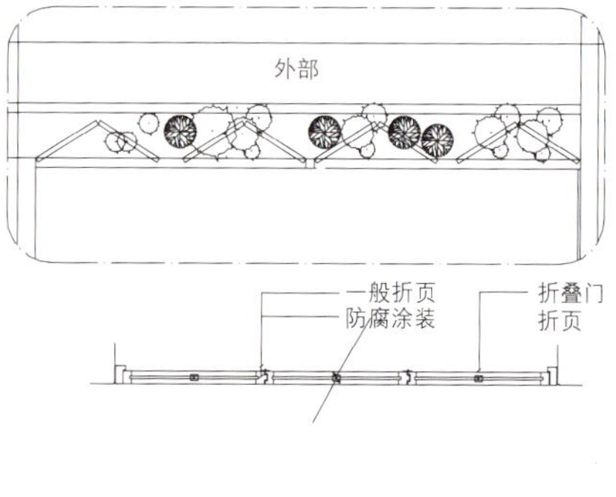

折叠门细部图

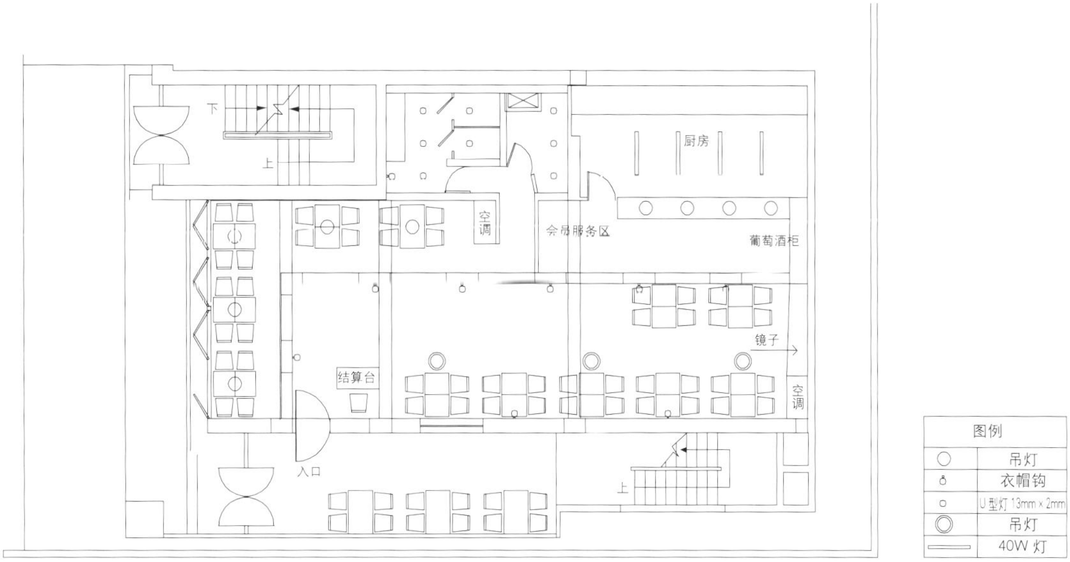

一层顶棚平面图

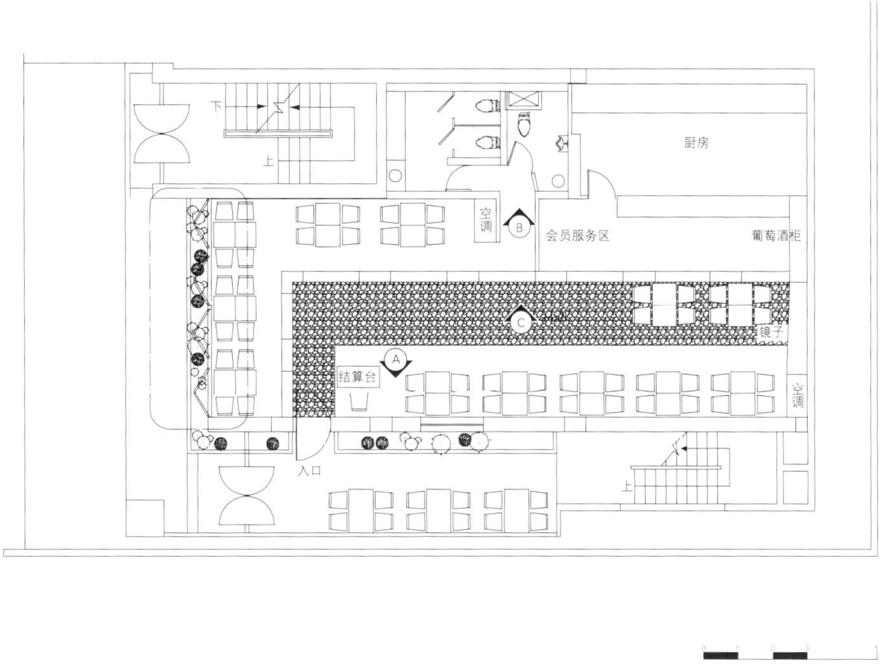

一层平面图

The client requested a considerably antique design for the legitimate Italian restaurant dealing in only Italian pasta and wine. The facade is made of a red brick wall for its old and aged feel. Especially, the sign puts the focus on maximization of the exoticism and antiqueness by applying corrosion coat on the timeworn materials found in junk shops in Songchu.

Based on the concept of a cave, the underground space creates a dramatic atmosphere through pumice stone furniture, the minimized illumination and partial lighting. 1st floor is the inside, but it takes a wide-open piazza as the concept and gives an image of an open-air cafe. The exotic lines of columns disappearing into the interior separate the space and would be regarded as a part of a square. The 2nd is an Italian wine bar. Because the wines have low popularity compared to its history and tradition, an old and antique atmosphere was exhibited as much as possible, thus giving an emphasis on its tradition. We aimed an unburdensome space with nobility, displaying naturalness through wood, fabric and stone.

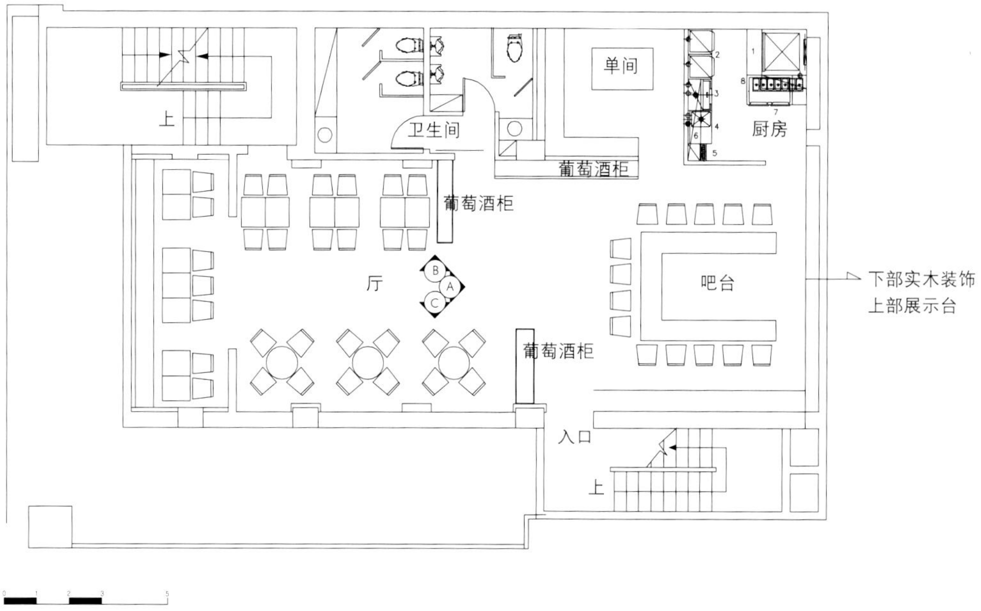

二层平面图

■ 二层餐厅

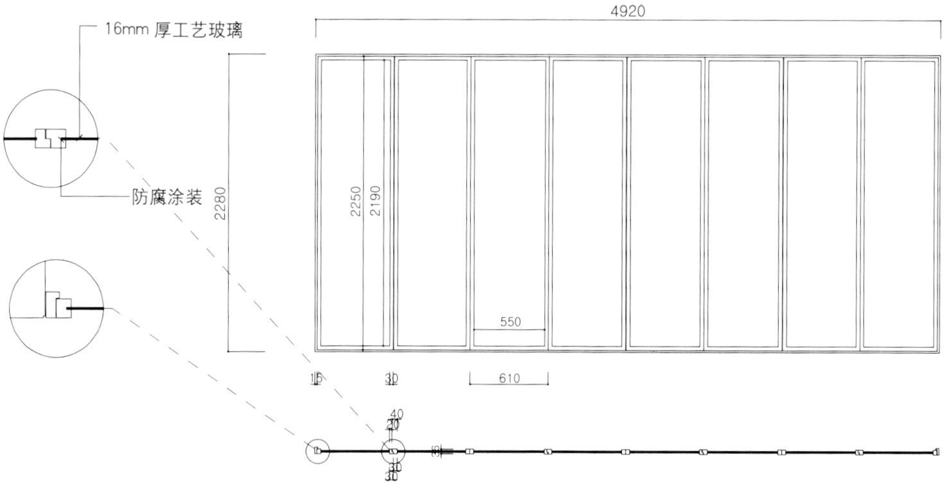

葡萄酒柜门—细部图

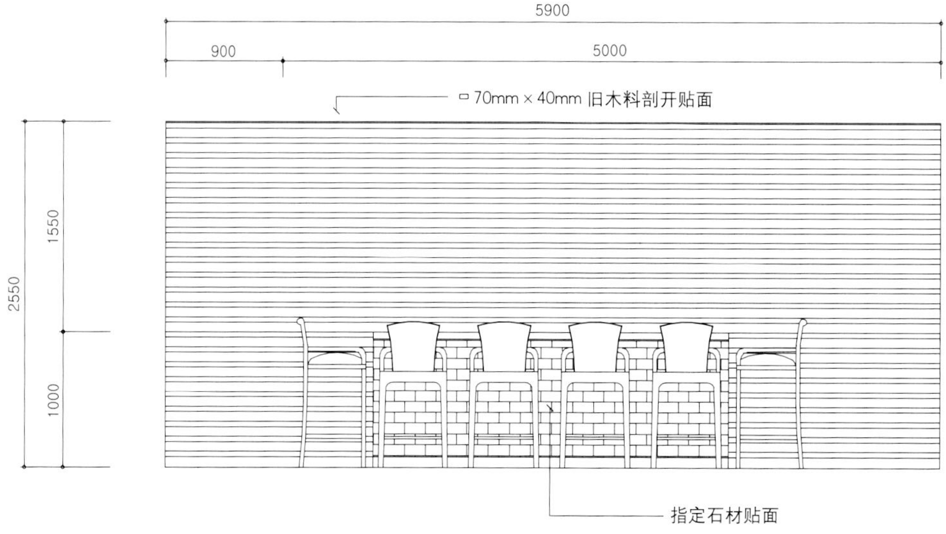

立面图 A

■ 葡萄酒柜

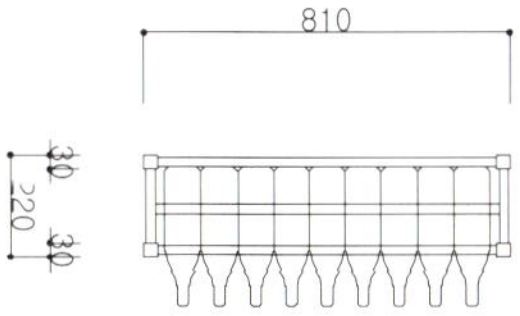

顶视图

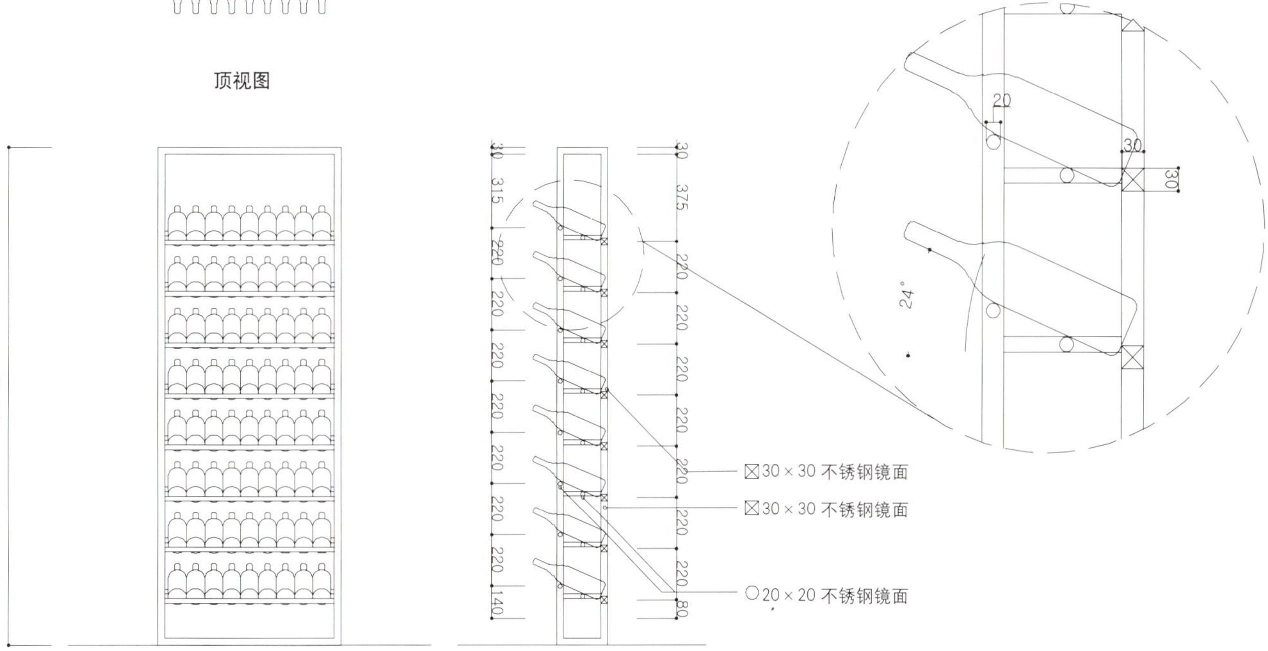

正面及剖面图（单位：mm）

■ 二层餐厅

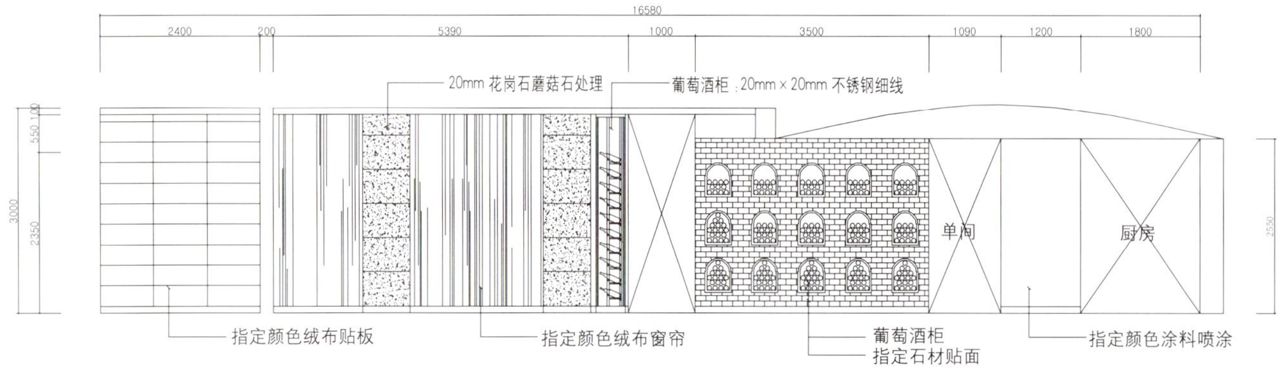

立面图 B

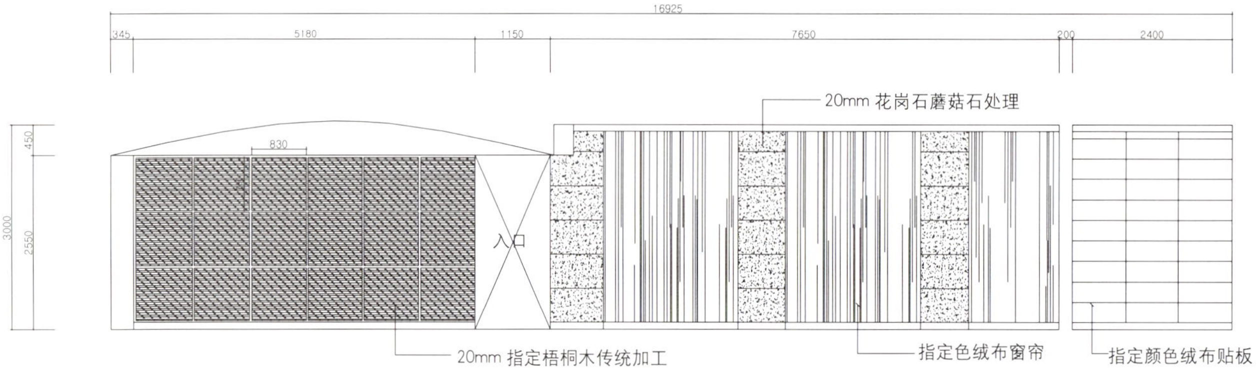

立面图 C

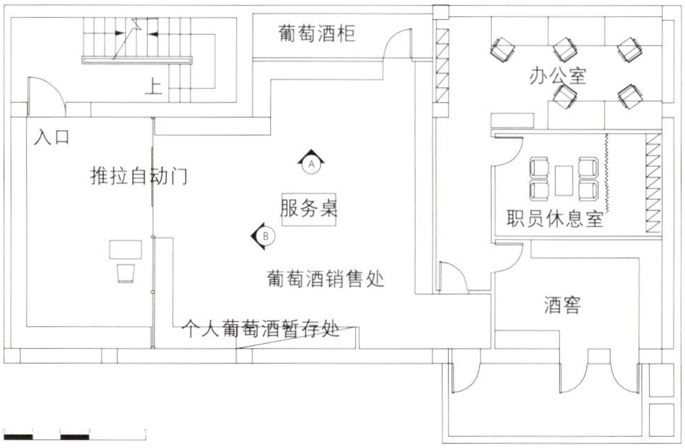

地下一层平面图

■ 地下一层葡萄酒销售

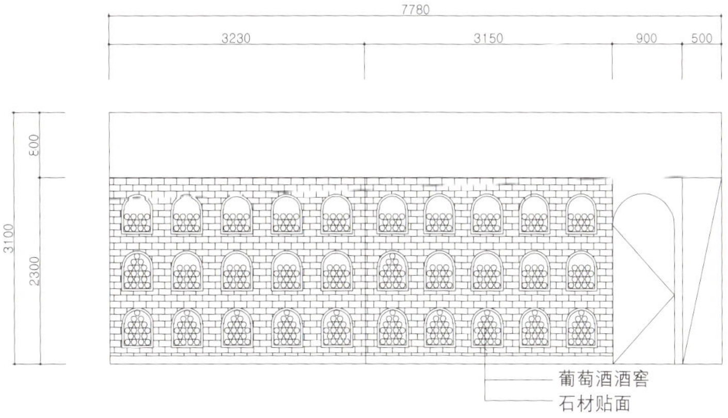

立面图 A

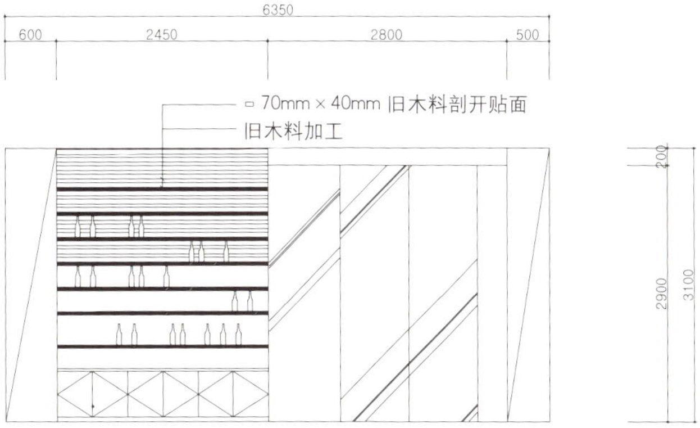

立面图 B

S—吧
纳芙迪葡萄酒吧

闵自助餐与酒吧
额尔古尔咖啡屋

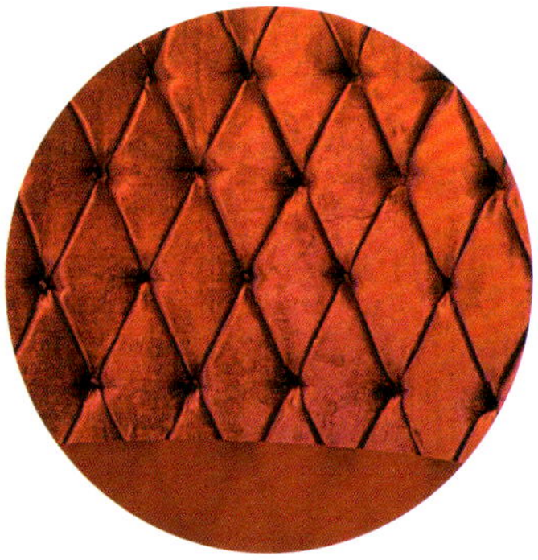

乐聚缘餐厅
翟氏咖啡屋

最新室内细部设计实例集 VI

自助餐厅与酒吧

Cafe & Bar

S-Bar

S－吧

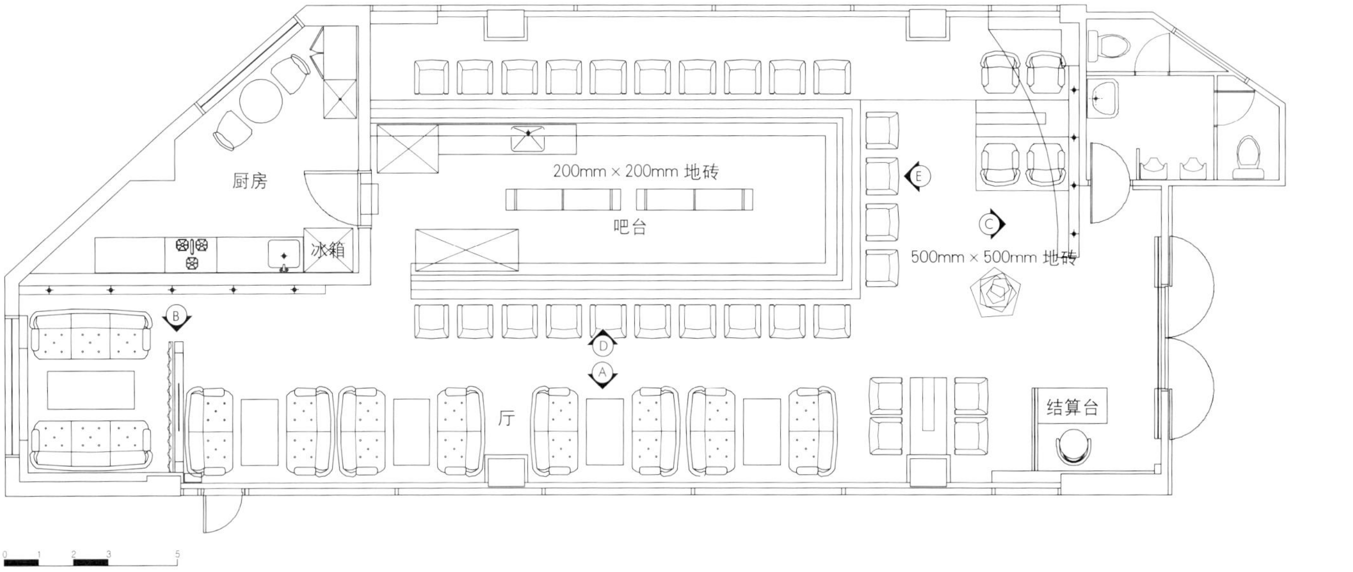

平面图

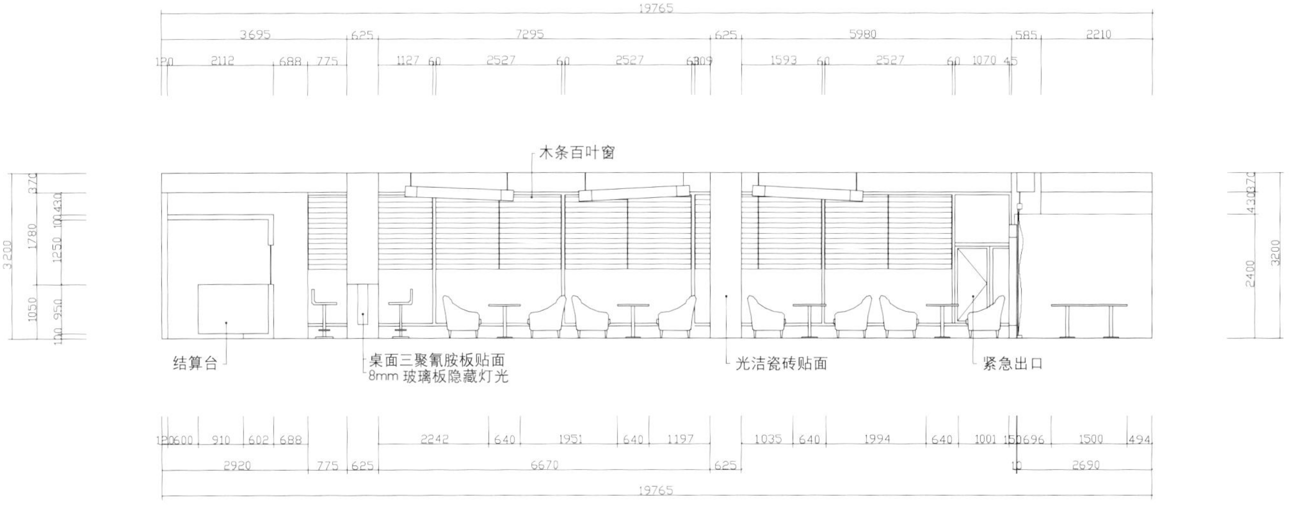

厅一立面图 A

这是条长长的隧道。以隧道的末端为起点多点相连组成了一个集合体——圆，设计师对圆形进行了延伸，以这种二维的形态作母题对空间进行了解析。就像经过隧道一样，圆形和四边形是非常有趣的。拥挤的空间中起着协调功能的设施——圆和四边形物件，同时也成了一处观光和视觉愉悦的产品。纤维编织物从顶棚悬落下来起着分隔空间的作用，从而，缓解了空间的紧张感，增加了空间的韵致。

以圆套圆的方式表现的“共存”概念，令人联想到祝福和祈祷，祝愿就餐的客人产生奇迹。用各种曲线进行的连接，表达了人们情绪上的某种变化。四边形赋予了空间以韵律并和圆环交织起来，在人和空间交融的圆形构筑物外，还有圆环，该圆环以外还有圆与圆的交叉重叠，如此这样一直循环下去。造型墙上花卉是叠放的，客人来此一一落座，不留空隙。他们谈论着、品尝着，在这里所有的物体都是相互关联的。

安静祥和！暖色调的天然木质薄板，不经意之间把空间的冷素材取代了。空间中温暖的沙发材质，以及椅子和桌子的尺度与人的视线的关系，将会给来此的客人留下不可磨灭的印象，耐人寻味。

设计：好友设计公司　梁镇荣（02）777—9914 | **施工：**好友设计公司 张荣权、赵成雄 | **地址：**首尔市东大门区全农 124—2 友成大厦 5 层 | **面积：**145.84m^2 | **装修：**地面／光洁瓷砖 墙壁／光洁瓷砖、6 号贴块、木纹木、彩色玻璃 顶棚／轻钢板材外露 上喷树脂涂料

Design: FRIEND'S DESIGN · Yang, Jin yeong I **Construction:** FRIEND'S DESIGN · Jang, Yeong gwon / Jo, Seong ung I **Built Area:** 145.84m^2 I **Finish:** Floor / Polishing Tile Wall / Polishing Tile, 6"Block, Veneer, Color Glass Ceiling / V.P over Slab Exposed

■ 结算台

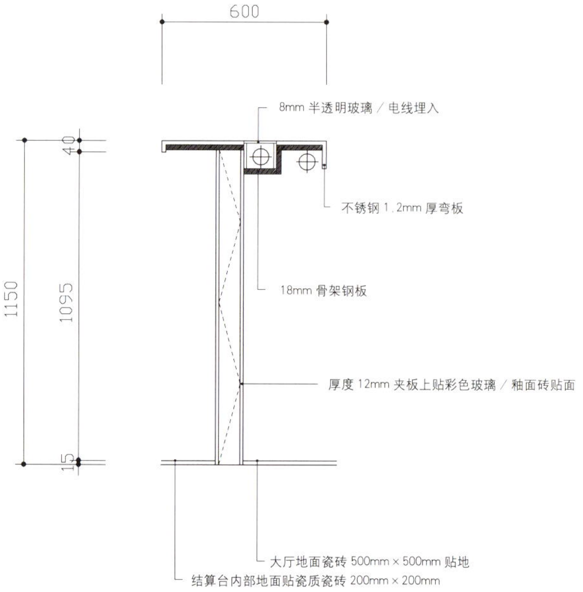

结算台剖面图

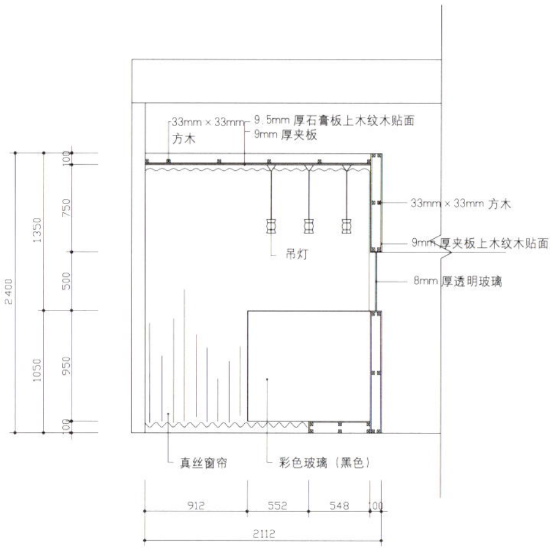

正立面图

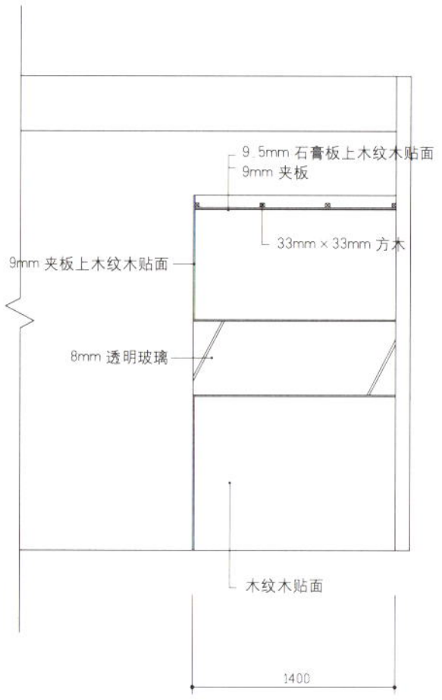

侧立面图

■ 厅

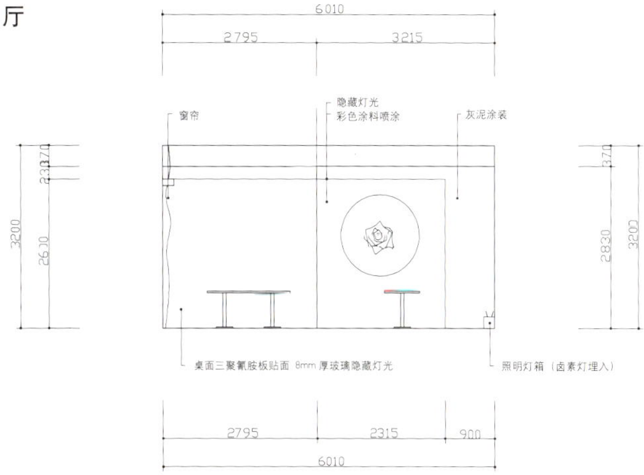

立面图 B

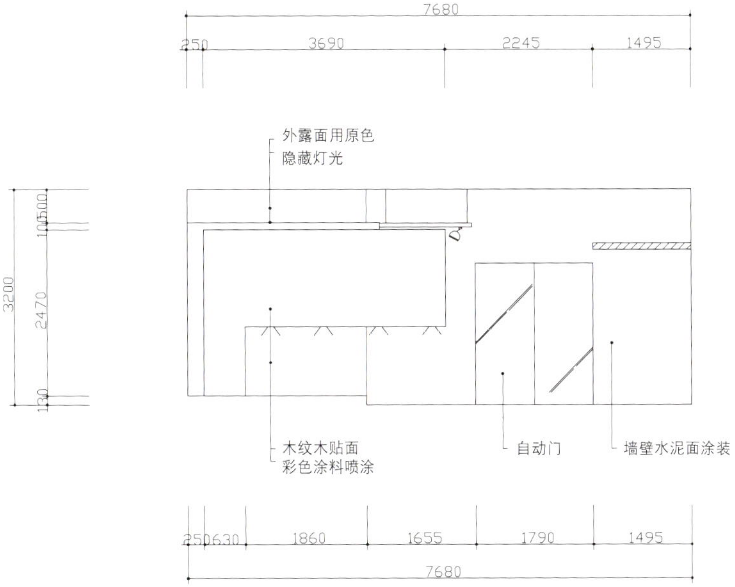

立面图 C

■ 桌子

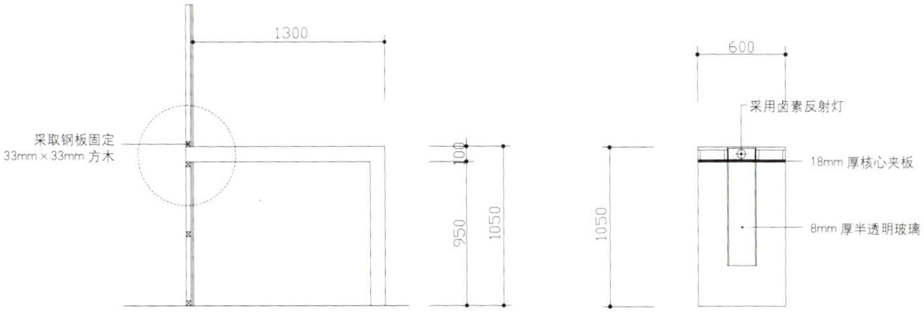

正立面图及顶视图

■ 吧台

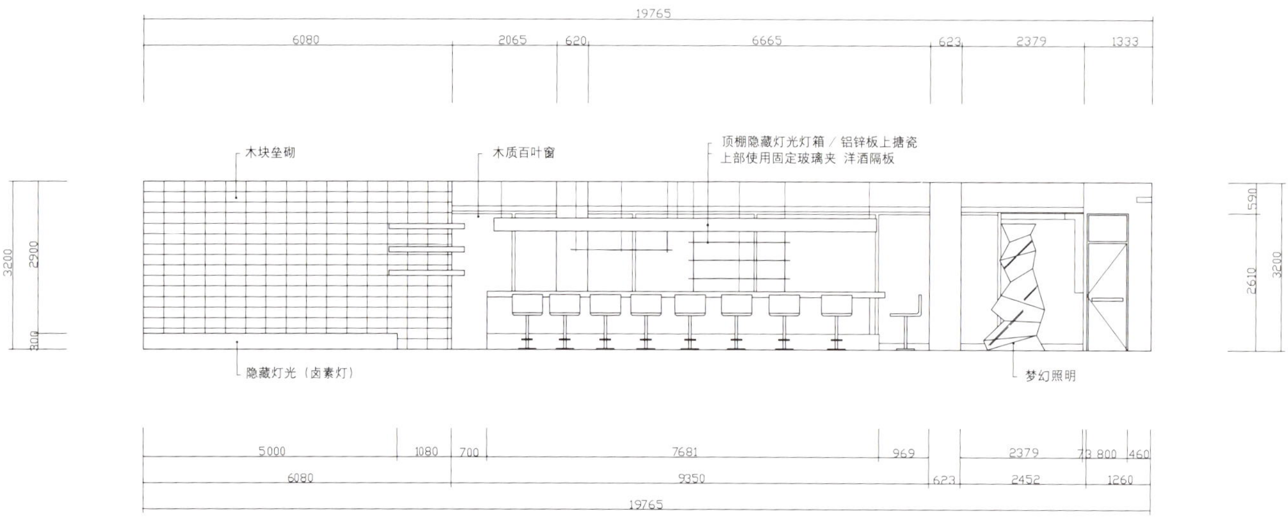

立面图 D

It is a long tunnel. The design motif lies on a two-dimensional surface beginning from the ending point of it, not just on a one-dimensional circular drawing. The roundness and squareness show such a fun, centering on a bar long as a tunnel. Objet comes to us as a visual amusement, being a device that filters a view and carries out space harmonization among the crowd. Fabric, which drops down from the ceiling for spatial division, relives the tension and evokes an artistic effect.

"Coexistence". I believe it is a result of a miraculous work achieved by small circles. Space and human being is unified as one through the naturalness resulting from invisibly connectivity created by curves, the repeated rhythm of squares balanced by circles, and ease of the tension. A round, structural net containing a circle with another circle inside it. Repetition of flowers on the image wall. People gradually occupy the seats and their stories fill the space. Everything is related.

Comfortable. Materials that could give a cold look are moderated by a natural-friendly wood blind, a warm material of sofa and the height and width of a chair and table, and the user's eye view - they form a entrancing space the visitors would never forget.

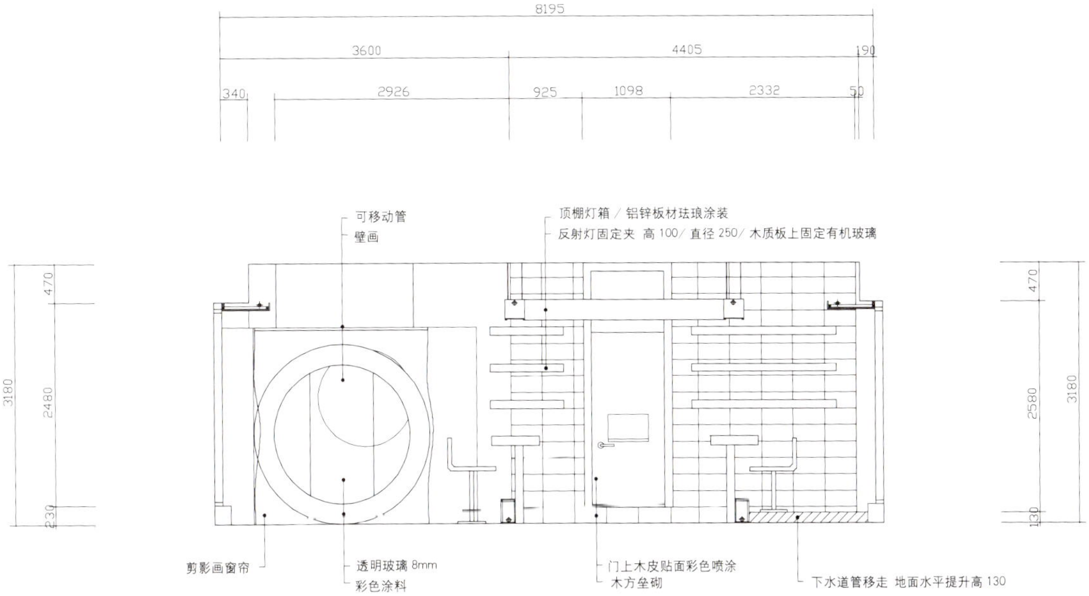
8195
3600
4405
190
340
2926
925
1098
2332
50
可移动管
壁画
顶棚灯箱／铝锌板材珐琅涂装
反射灯固定夹 高 100／直径 250／木质板上固定有机玻璃
3180
2480
470
230
470
2580
3180
130
剪影画窗帘
透明玻璃 8mm
彩色涂料
门上木皮贴面彩色喷涂
木方垒砌
下水道管移走 地面水平提升高 130

立面图 E（单位：mm）

Lapodo

纳芙迪葡萄酒吧

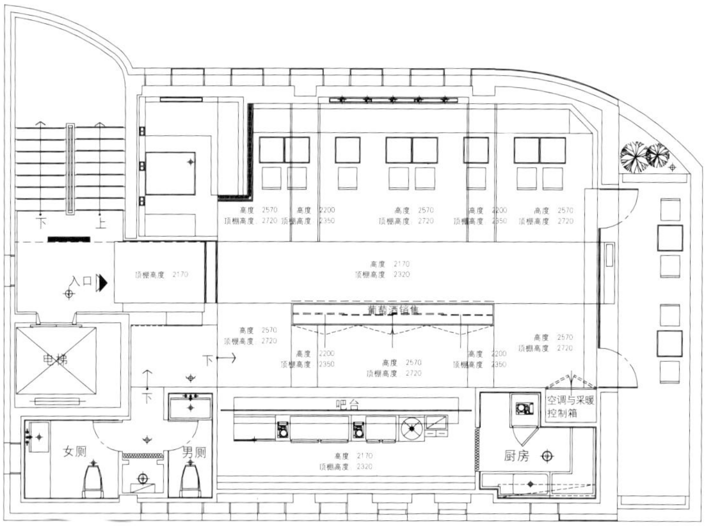

顶棚平面图

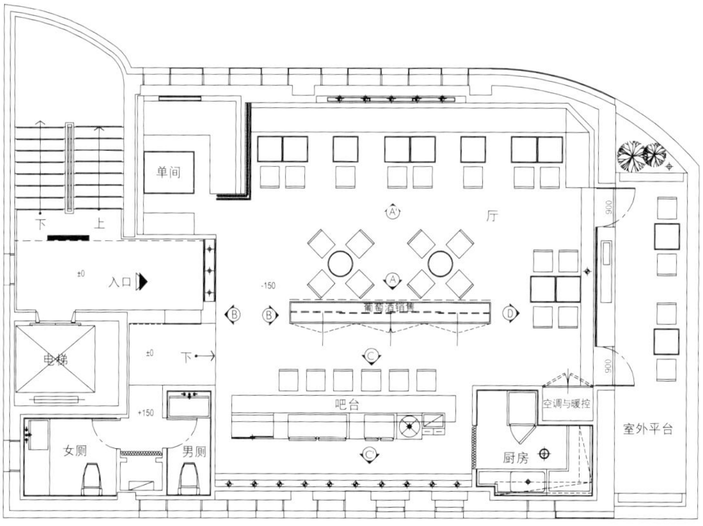

平面图

葡萄酒时尚的流行起始于首尔市江南区的罗代奥大街，就像丰盛的葡萄酒宴一样，虽然不是很醒目，但是却非常有深度，这就是“纳芙迪葡萄酒吧”。走过嘈杂喧闹的小巷，典雅别致的建筑物就坐落在小巷的尽端高地上。酒吧位于该建筑的五层，站在酒吧的入口处看到的是珠帘后面镜面顶棚一直延伸到墙边，一株很粗的大树树干就安放在那里，大树一直从地板伸到顶棚，就像是空荡街道上的红色葡萄酒，为酒吧增色不少。

顶棚上的镜子使树木看起来更加庄重挺拔，树干也使镜子看起来更加柔和亲切。这两种材料的默契配合，与其说增加了空间的现代感，毋宁说使客人回归自然。

一般来说，葡萄酒吧应该避开强光的照射。所以，本酒吧的大窗户不是用来自然采光，而是设计成可以接受“黑暗光”，空间看起来也更有某种神秘的现代感。

每处入口的主题珠帘都有各自的图案，它本身形成一个层，这个层强化了贯通中厅的镜面轴线。位于大厅中央的葡萄酒销售点不仅仅位于中轴线上，而且也起到划分空间的作用。从传送带之间看到的排列整齐的葡萄酒瓶更加深了空间的深度。酒吧和包房内的珠帘和墙纸作为一种中间层，在珠帘和墙纸之间的光线也形成了一层空间。

概念设计：（株）EOE 设计公司　金光林（02）2057−8202 |**设计：**（株）EOE 设计公司　吴镇成 |**施工：**（株）EOE 设计公司　金永民 |**地址：**首尔市江南区新沙洞 663−11 银山大厦五层 |**面积：**165m² |**装修：**地面／人造大理石、印茄木地板　墙面／涂装、钢化玻璃、壁纸　顶棚／涂装、镜子 |**合作公司：**三星涂装、大申玻璃、陶瓷石材 C&M、友谊金属、新荣阳台

Basic Design: eoedesign associates · Kim, Kwang lim I **Design:** eoedesign associates · Oh, Jin sung I **Construction:** eoedesign associates · Kim, Young min I **Built Area:** 165m² I **Finish:** Floor / Artificial Marble, Merbau Wall / V.P, High Glass, Wallpaper Ceiling / V.P, Silver Mirror

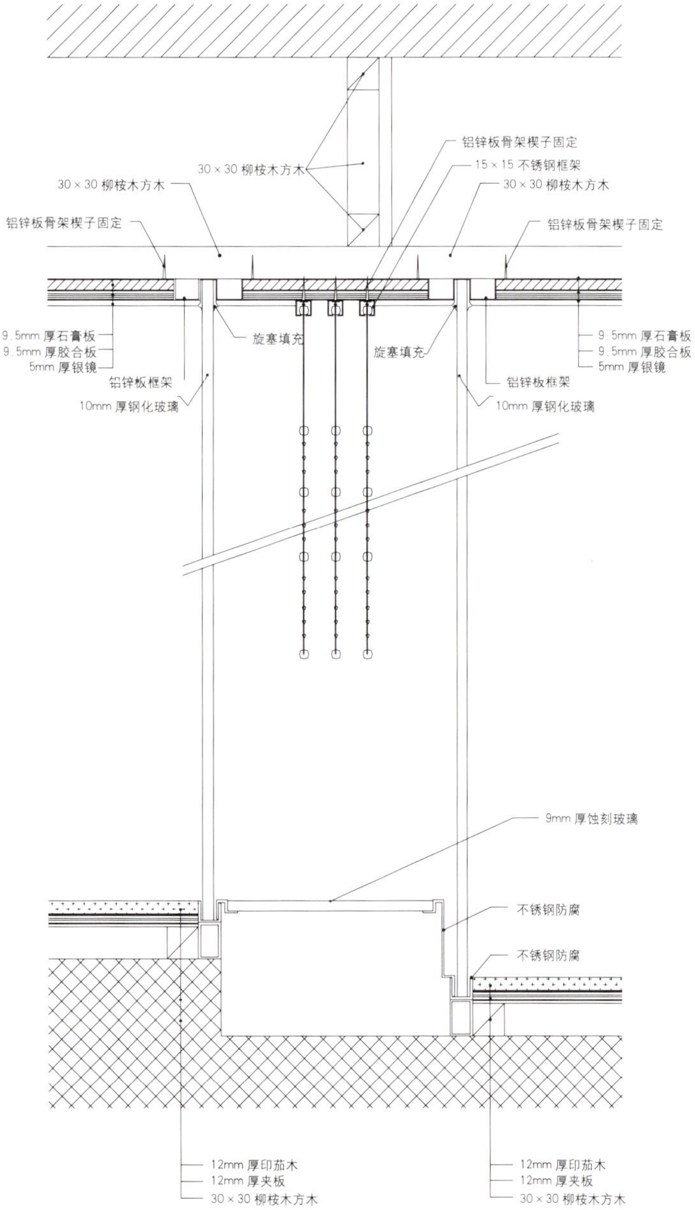
铝锌板骨架楔子固定
15×15 不锈钢框架
30×30 柳桉木方木
30×30 柳桉木方木
30×30 柳桉木方木
铝锌板骨架楔子固定
铝锌板骨架楔子固定
9.5mm 厚石膏板
9.5mm 厚胶合板
5mm 厚银镜
旋塞填充
旋塞填充
9.5mm 厚石膏板
9.5mm 厚胶合板
5mm 厚银镜
铝锌板框架
10mm 厚钢化玻璃
铝锌板框架
10mm 厚钢化玻璃
9mm 厚蚀刻玻璃
不锈钢防腐
不锈钢防腐
12mm 厚印茄木
12mm 厚夹板
30×30 柳桉木方木
12mm 厚印茄木
12mm 厚夹板
30×30 柳桉木方木

入口剖面（单位：mm）

■ 厅

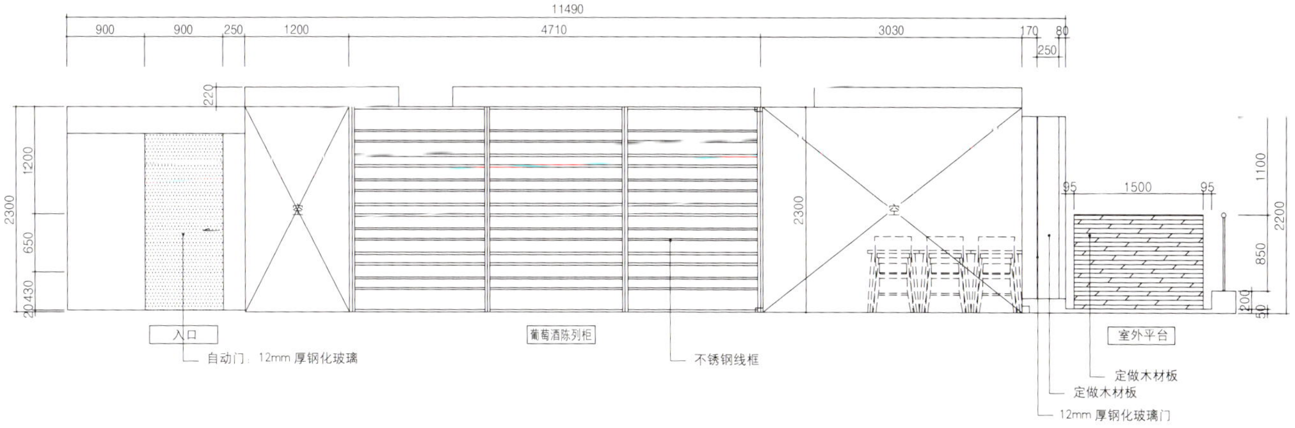

11490
900
900
250
1200
4710
3030
170
80
250
220
1200
2300
650
20
430
空
2300
95
1500
95
1100
2200
850
200
50
入口
自动门：12mm 厚钢化玻璃
葡萄酒陈列柜
不锈钢线框
室外平台
定做木材板
定做木材板
12mm 厚钢化玻璃门

立面图 A

■ 厅

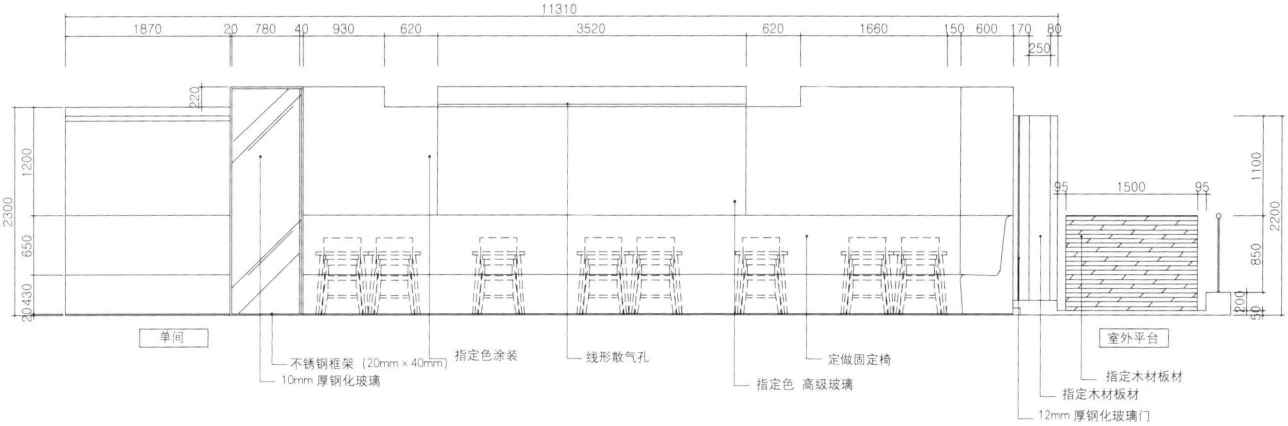

立面图 A’

Like wine on a generous table, though without a biting feel, Lapodo is a profound space for fanatics on Rodeo street, the starting point of fashion. Outside a narrow street full of chaos and magnificence, you can see Lapodo on top of a tidy building. Once inside, standing beyond the beads at the end of a silver mirror is a tree, as enchanting as red wine. Silver mirror gives majestic beauty to the tree that softens its coldness. Owing to mutual support of the two materials, naturalness is added to a modern space.

Since a wine bar needs darkness, a large window of Lapodo accepts dark instead of light and creates a shadow joined with an illumination in an overall modern space. Objet beads in the entrance become a pattern and the surfaces they make create a single layer, emphasizing the axis toward the silver mirror. The wine cellar on the center is a central axis and a partition that separates the space. Seen between conveyor belts, the wine bottles, which are arranged in a pattern, add depth to the two spaces. In a bar and room, a layer is created from reiteration of a ball chains and wall, and another layer from the light going across the two materials.

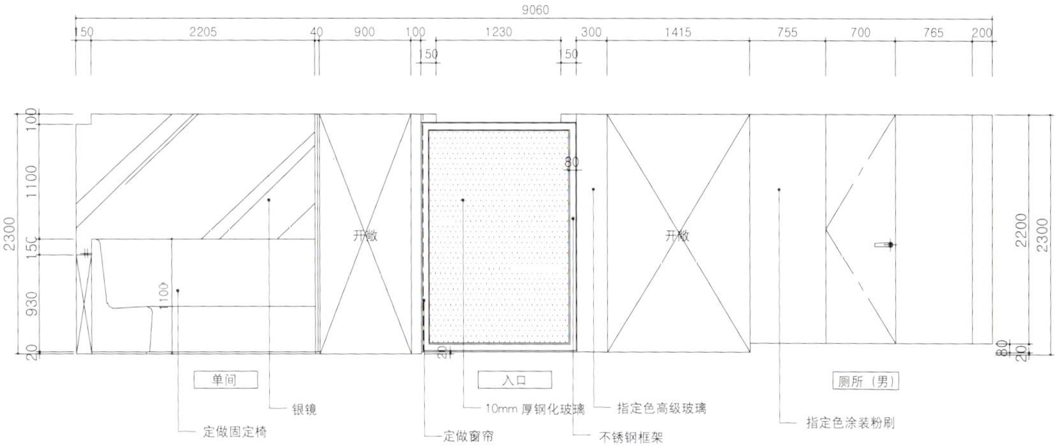

开敞
开敞
单间
入口
厕所（男）
银镜
定做固定椅
10mm 厚钢化玻璃
定做窗帘
指定色高级玻璃
不锈钢框架
指定色涂装粉刷

立面图 B

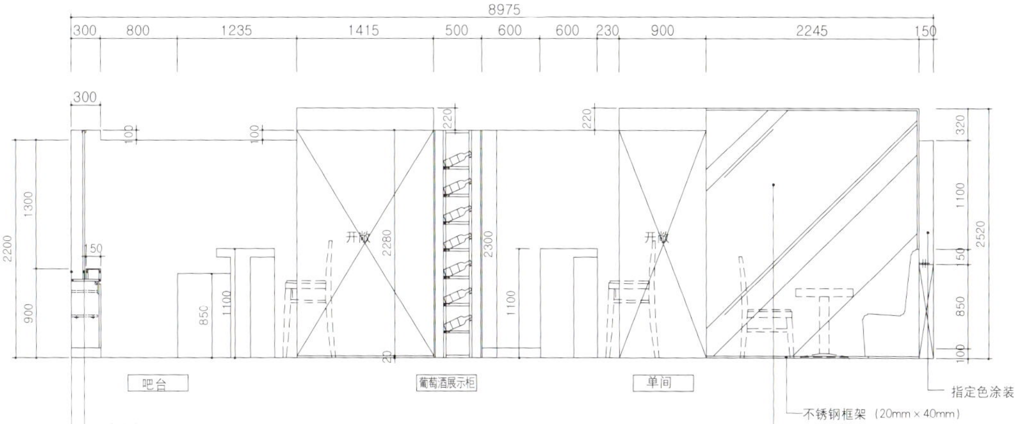

开敞
开敞
吧台
葡萄酒展示柜
单间
指定色涂装
不锈钢框架（20mm × 40mm）
10mm 厚钢化玻璃
珠子串
指定壁纸

立面图 B'

■ 吧台

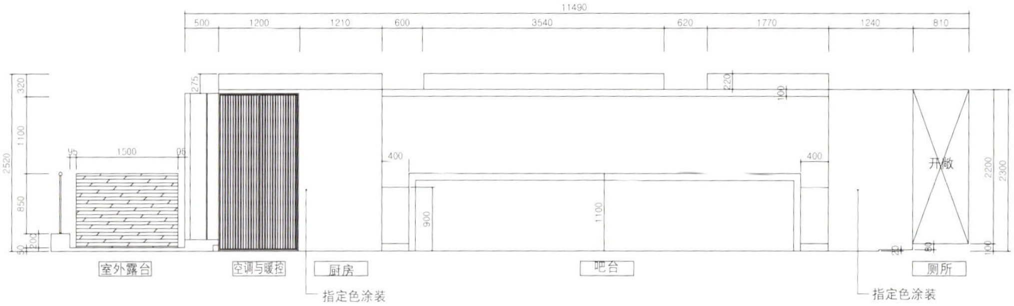

立面图 C

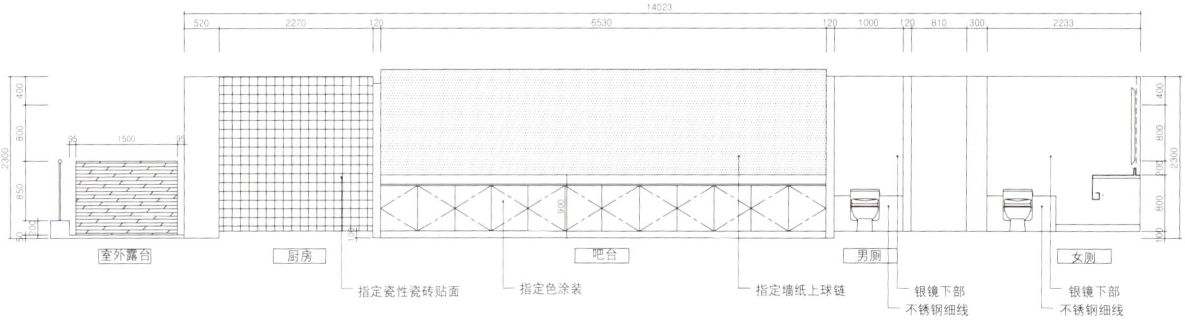

立面图 C'

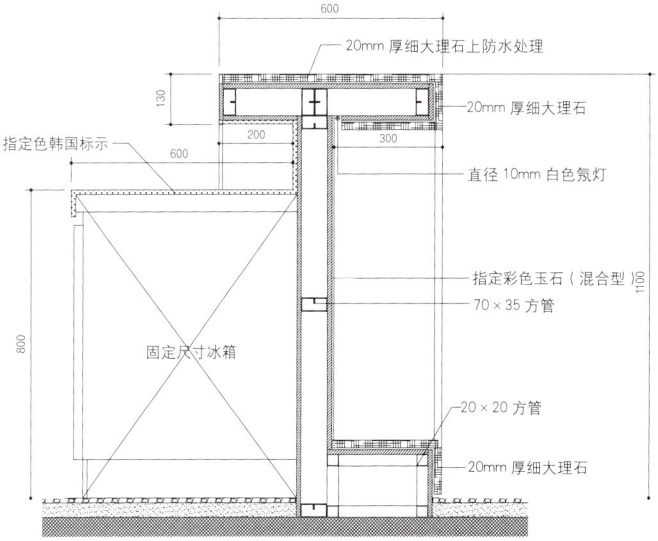

剖面图（单位：mm）

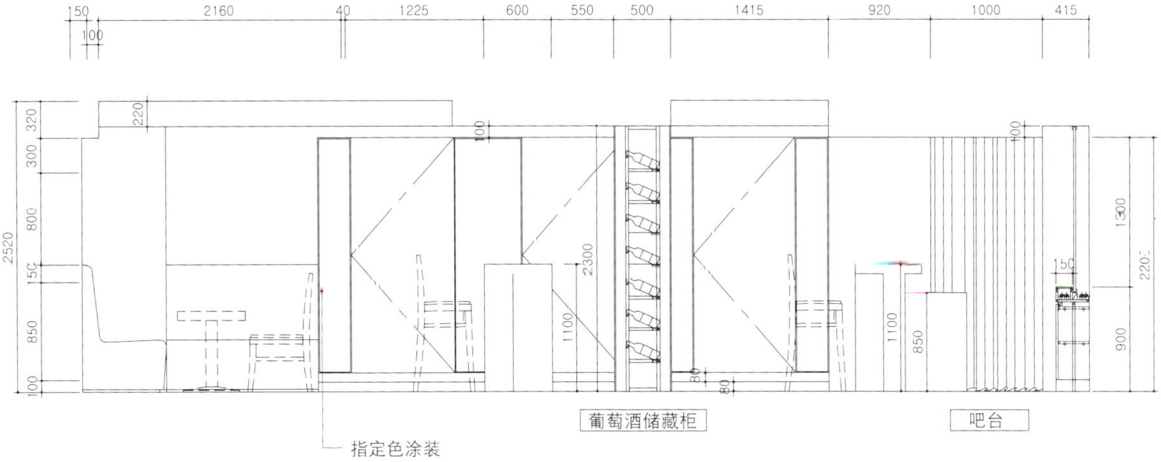

立面图 D

细部图

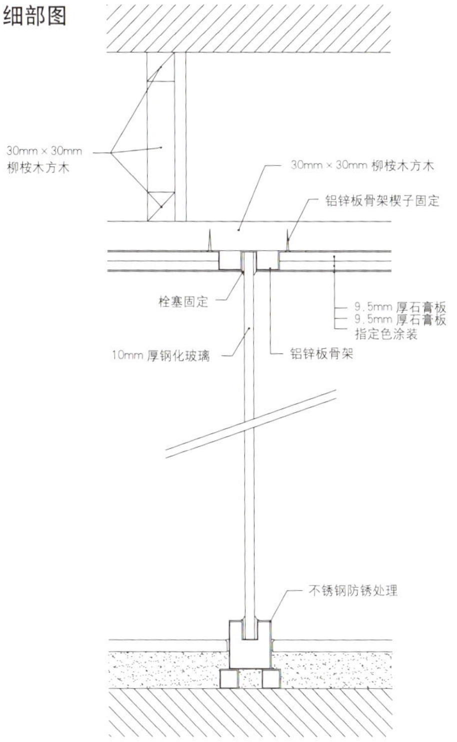

单间隔断剖面图

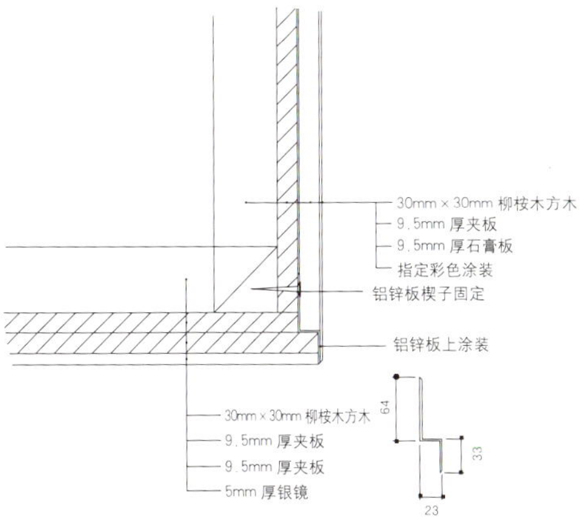

中心顶棚玻璃板剖面图

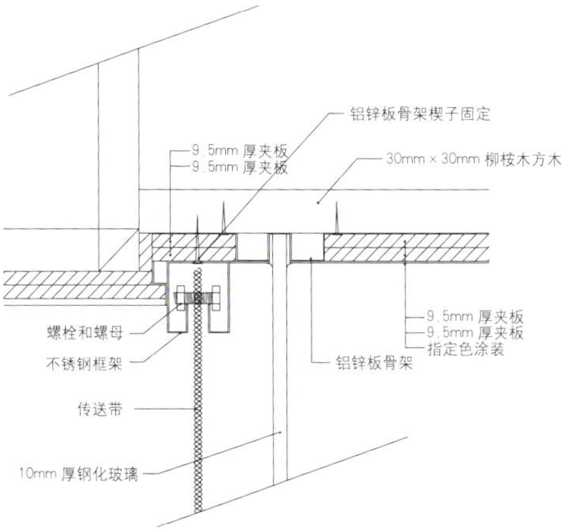

中心传送带剖面图

Min Cafe & Bar

闵自助餐与酒吧

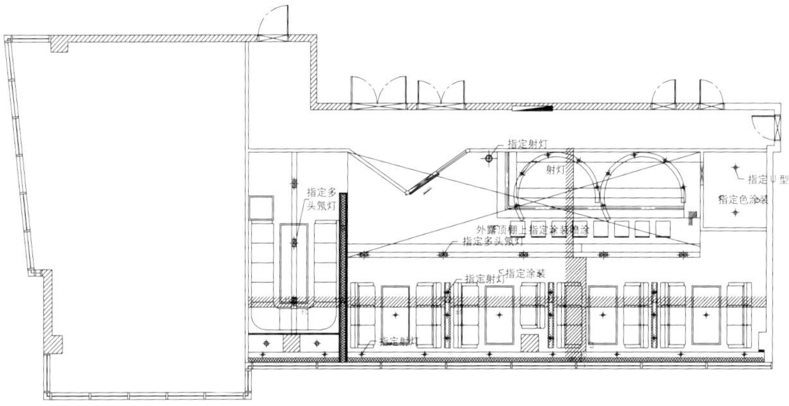

顶棚平面

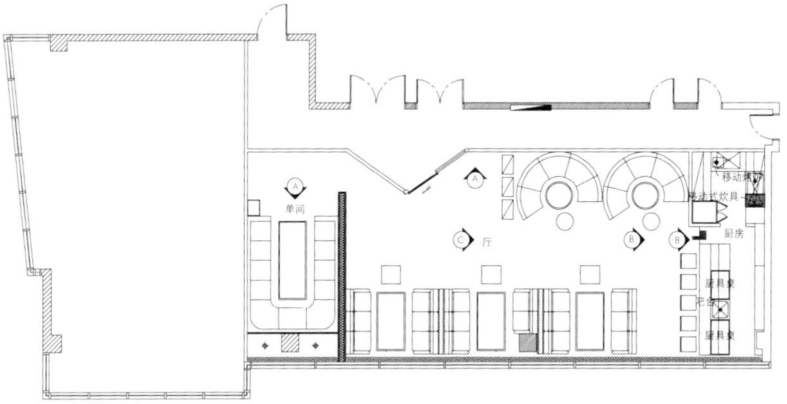

平面图

站在新建建筑的二层走廊内，与建筑的出入口相对应的地方，有一处凹进去了，那里就是酒吧的入口大门。该门的布局非常特别，不仅是酒吧的入口也是一面艺术墙。

推开大门进入厅内，一团红色映入眼帘，气氛非常安静。大厅的一侧用红色窗帘装饰，烘托出时尚空间，给人一种别样的感觉。就像在看宗教电影一样，空间的色彩感更加深了一层。

酒吧的包间隔断采用的是编织物和玻璃，包间内的各项自然元素反复排列使用，从而引起了空间的变化。室内大部分采用局部照明，每个物件都作为空间中的亮点，在室内交替出现，形成了另外一个层。大厅内的各个层之间也是相互联系的。

形成金属层的大空间由若干个部分组成，每部分又分别占据着各自的空间。矮的屏风遮挡了下部视线，却使上部能够通览整个大厅，从而把室内的色彩统一了起来。红色的窗帘对面安排的是沙发，沙发的颜色更加鲜艳。当然，沙发也是一个层，它也在同一个色系中。漫步走向吧台可以看到简约设计的图案化艺术墙，墙上的顶棚与此非常协调。

On the second floor of the new building is an entrance designed in the extreme moderation. You can also see it as an art wall of the hall.

Inside, intense red tone blends into an overall calm atmosphere. A wall finished in red curtains contributes a different feeling to the modern space. Colors, which make you feel as if in a cult movie, elevate the bar's luxuriousness.

Fabric and glass layers divide the rooms, and their natural elements are displayed with different feelings through alteration and repetition of layers. Partial illumination was mainly used, and accompanying accent lightings were overlapped to the overall color, forming yet another layer. Layers are continued in the hall. All booths made of metal layers make a single space and, at the same time, stand alone as individual spaces. Colors are unified by a partition, which blocks the sight provides a whole view at a glace. Facing the impressive red curtain is a sofa with stronger intensity. A layer is also found in the sofa. Though in same color, it has a different atmosphere from that of the hall. On a bar side, you will find a hung-down art wall in simple design and the ceiling on which it is attached connects the whole space.

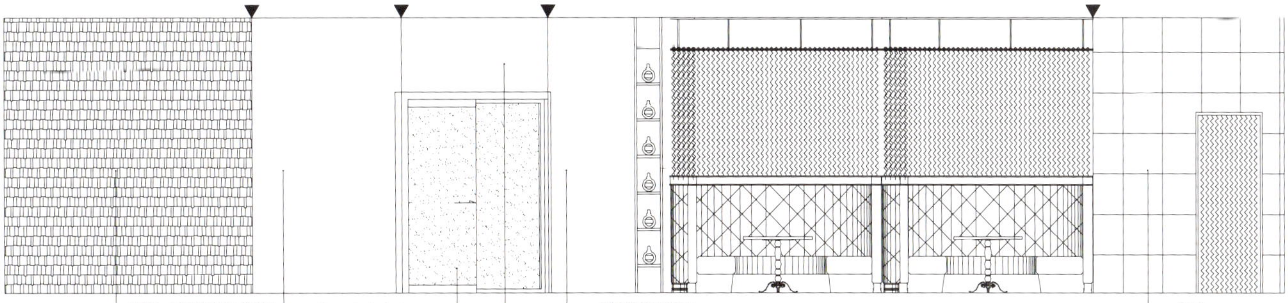

厅一立面图 A

概念设计：（株）EOE 设计公司 金光林 （02）2057–8202| **设计：**（株）EOE 设计公司 崔允正 | **施工：**（株）EOE 设计公司 文昌秀 | **地址：** 首尔市江南区浦一洞 | **面积：** 104m^2 | **装修：** 地面 / 指定瓷砖 墙面 / 涂装、纤维编织壁纸、窗帘、玻璃上白色喷漆 顶棚 / 涂装

Basic Design: eoedesign associates · Kim, Kwang lim I **Design:** eoedesign associates · Choi, Yoon jung I **Construction:** eoedesign associates · Moon, Chang soo I **Built Area:** 104m^2 I **Finish:** Floor / App. Tile Wall / V.P, Twine Wallpaper, Curtain, Back Painting Glass Ceiling / V.P

■ 厅

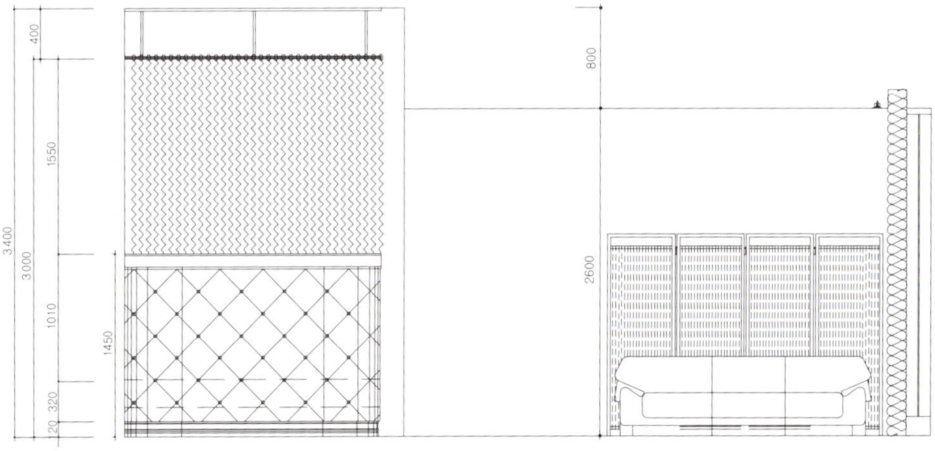
3400
3000
400
1550
1010
320
20
1450
800
2600

立面图 B

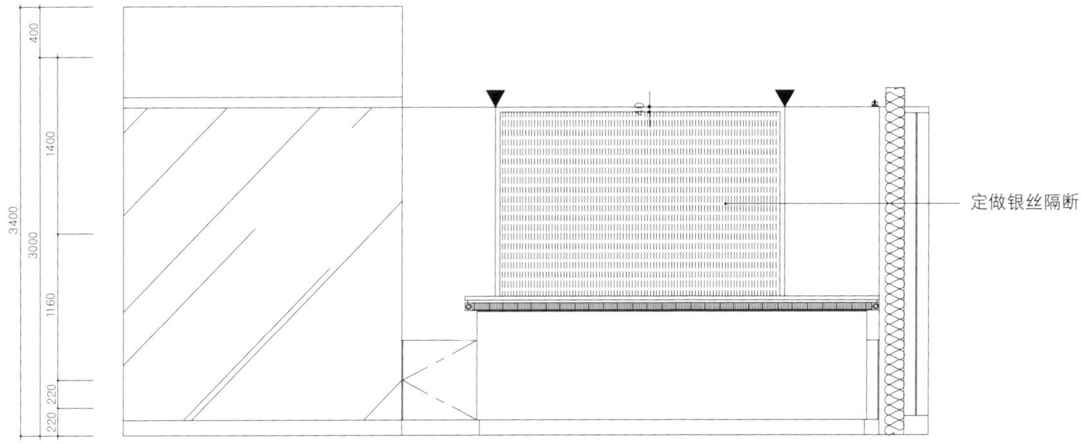
3400
3000
400
1400
1160
220
220
定做银丝隔断

立面图 B'

■ 墙面

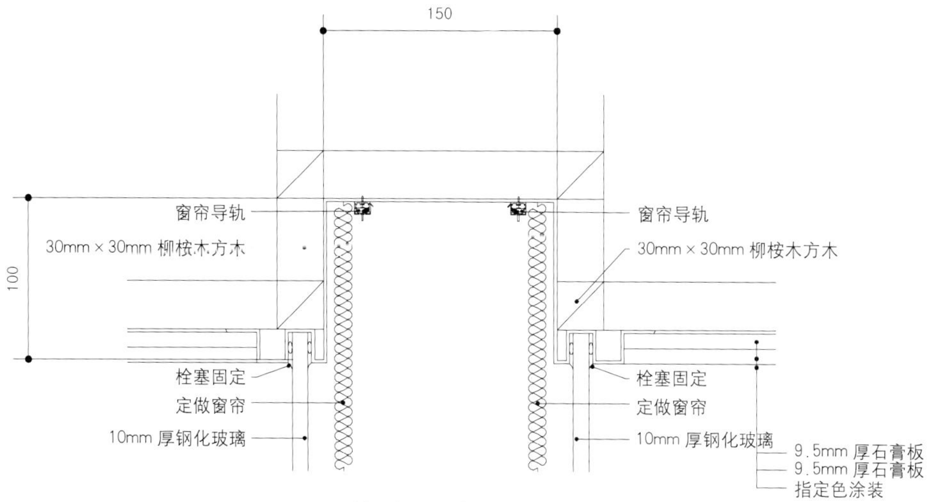

单间和厅一墙壁剖面图

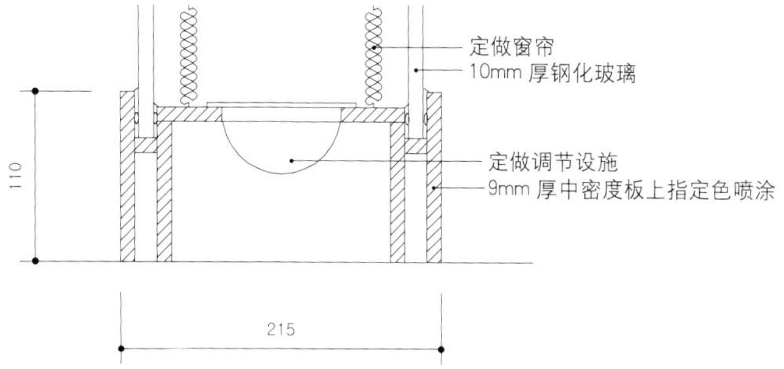

厅一墙壁剖面图

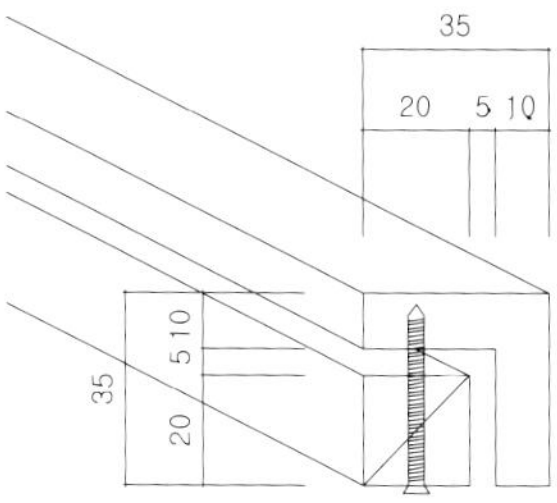

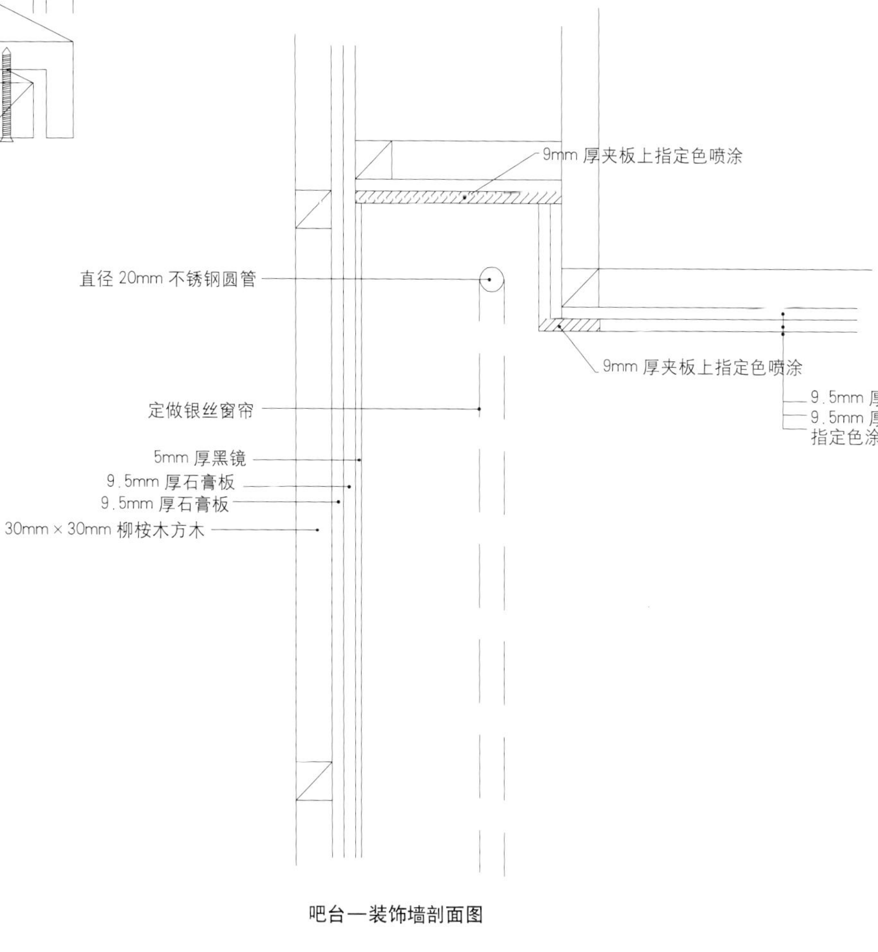

吧台一装饰墙剖面图

■ 厅

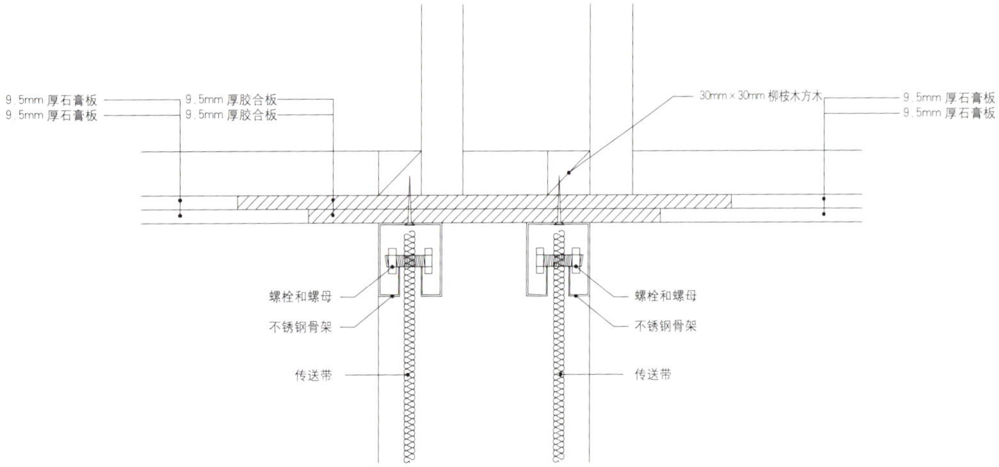

隔断一剖面图

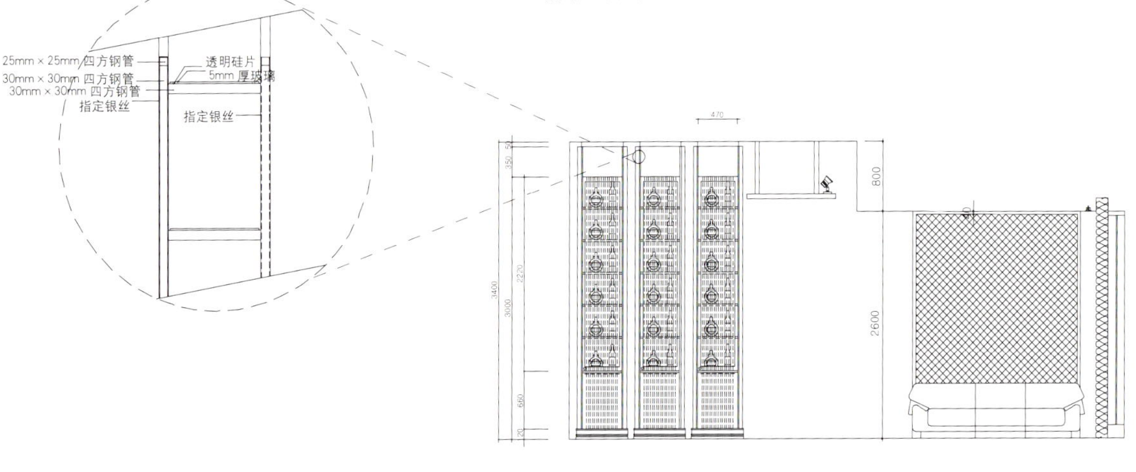

立面图 C

■ 单间

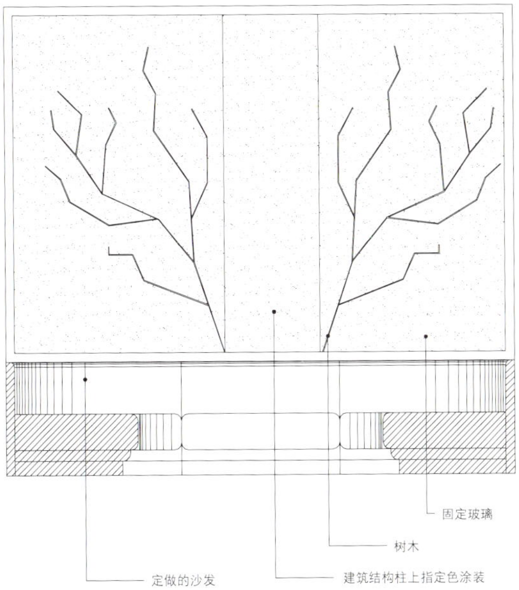

定做的沙发
建筑结构柱上指定色涂装
树木
固定玻璃

立面图 A

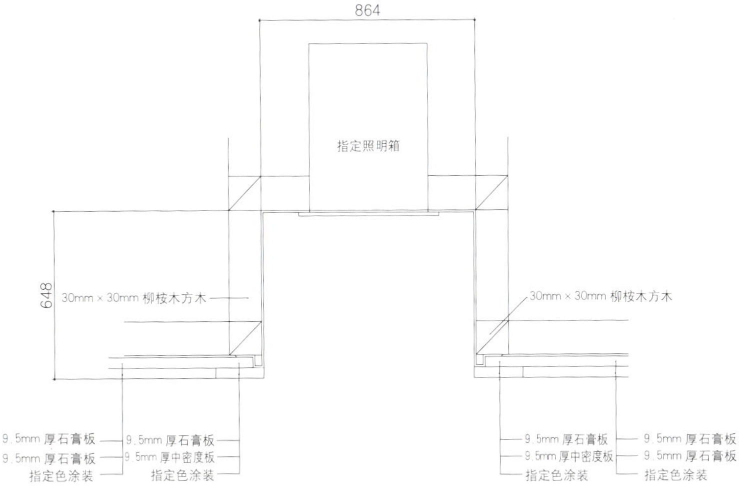

864
指定照明箱
648
30mm × 30mm 柳桉木方木
30mm × 30mm 柳桉木方木
9.5mm 厚石膏板
9.5mm 厚石膏板
指定色涂装
9.5mm 厚石膏板
9.5mm 厚中密度板
指定色涂装
9.5mm 厚石膏板
9.5mm 厚中密度板
指定色涂装
9.5mm 厚石膏板
9.5mm 厚石膏板
指定色涂装

灯箱

Cafe Eulgul

额尔古尔咖啡屋*

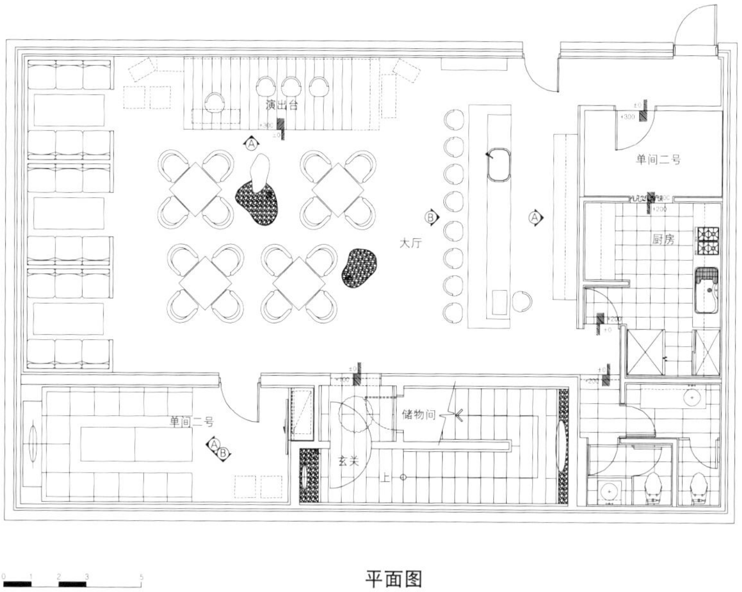

平面图

“画圆不成画成脸……”这是一首20世纪70年代的校园歌曲，歌曲的名字就是“脸”。这首歌是流行歌手允元贤演唱的，最近，她邀请喜欢乡村音乐的朋友开办了一个小型娱乐空间，空间的名字就是额尔古尔，该空间位于首尔市西桥洞。人们在这里可以唱歌回忆过去的时光，也可以迎接美好的未来。

站在楼梯间入口处，向上看有很多大小厚薄不匀的木块贴在墙上，整体看起来虽然很简单，但立体感非常强，有的地方较粗糙，有的地方动感强烈。设计中把这些木板放到了上部，在下面预留了空间，隐藏灯光从上面直接照射下来，强化了入口。进入大厅最能吸引眼球的是一直延伸到顶棚的绿树。如果说木板是间接地表达自然感觉的话，绿树就是直接表达自然的感觉。这样的布置引起人们对乡村音乐的怀念，同时也表达了乡村音乐悠长的生命力。就如同树木一样根深叶茂……

一侧墙面上用LP板材密密麻麻排列的是老歌手宣传画，室内飘过的木炭味，使人对往事浮想联翩。看到这些曾经出名的或者尚未出名的乡村歌手的熟悉面孔，人们仿佛回到了过去，看到了自己的影子，也想起了同时代的这些歌手。最初的墙面具有了找寻记忆并回味过去的作用。

主要的建筑装饰材料是木材和韩宣纸。木材具有天然的质感，韩宣纸具有温暖的质感，这两者的生命力是非常接近的。几层宣纸叠在一起硬度和厚度适中，再与小块木片结合起来的搭配不仅坚固耐用还有古朴的感觉，达到整体空间的布局统一，到此就算大功告成了。

当你在工作生活中经常碰壁。到哪里去找一处休息地呢？当你想与老友一起回忆往日时光时，当你打算欣赏乡村音乐的曲调时，何不在额尔古尔咖啡屋中体验一下呢？

‘I drew a circle and find out I actually drew a face...’ Eulgul(face), a song anybody must have sung more or less once in his school days. Yun Yeon Seon, the singer of the song, has arranged a space that she can share with other folk lovers. There, people would reminisce the old days of folk and fill the coming days with it also.

At the entrance of a staircase, woodblocks of irregular size and thickness cover a long wall. The overall surface is simple but has a cubic effect to give a rough, and sometimes vital look. An indirect lighting brings warmth to the entrance. The most remarkable things in the hall are the trees that touch the ceiling, full in the space. Woodblock indirectly expresses nature where the trees take a direct approach, drawing it into the hall. This increases nostalgia for folk music and also means a long life of the genre. Just like a straight wood branch that wouldn’t break or wither.

A wall of LPs leads us to fall deep in thoughts, as the smell of burnt charcoals brings back the old memories. The familiar faces of present or former folk singers remind people of themselves and their friends in those days. Then, the walls function as a device of remembrance and recollection.

Woodblock and Korean paper are the main materials. It seems that life rests on the natural textured woodblock and warm Korean paper. Layers and layers of paper and a series of woodblock lend a solid and aged feel. Here one can breathe out a deep breath.

When in need of a resting place being blocked against the wall of reality, when in want of recollection of the old days with old friends, when in demand of enjoyment through folk music, how about taking five in 'Eulgul'?

* 额尔古尔在韩语里是“脸”的意思，来自于韩国歌手允元贤（1953年生）20世纪70年代所演唱的一首歌曲，大意是画圆不小心画成了相思人的脸，比喻坠入情网的情侣。——译者注

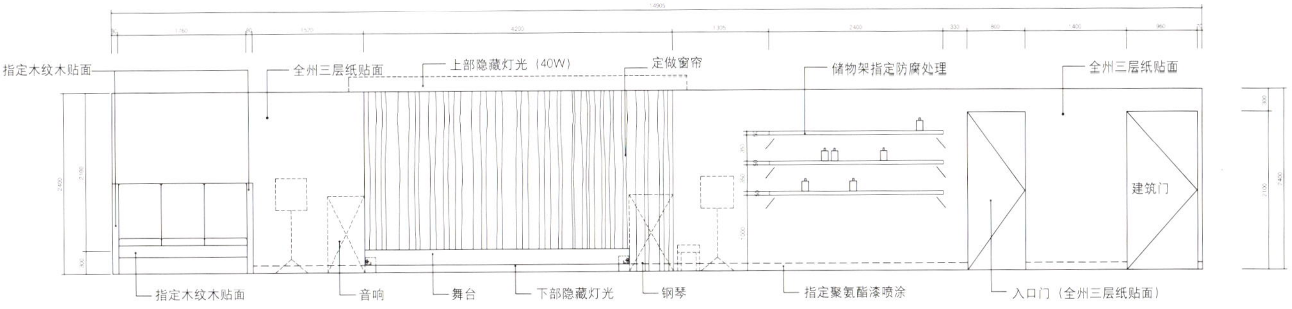

大厅一立面图 A

设计：核心手室内设计公司 刘在峰 李光烈（02）396—2845 | **地址：**首尔市麻浦区西桥洞 | **面积：**138.74m² | **装修：**地面 / 釉面砖、实木地板、碎石子、瓷砖 墙面 / 圆木、木块、全州三层纸、黄土内墙腻子、灰墙、防腐蚀涂装 顶棚 / 树脂喷涂

Design: COREhands Interior Design · Yu, Jae bong / Lee, Gwang yeol | **Built Area:** 138.74m² | **Finish:** Floor / P-Tile, Wood, Pea Gravel, Tile Wall / Material Wood, Wood Block, Korean Paper, Loess Plaster, Plasterd Wall, Corrsion Painting Ceiling / V.P

■ 吧台及结算台

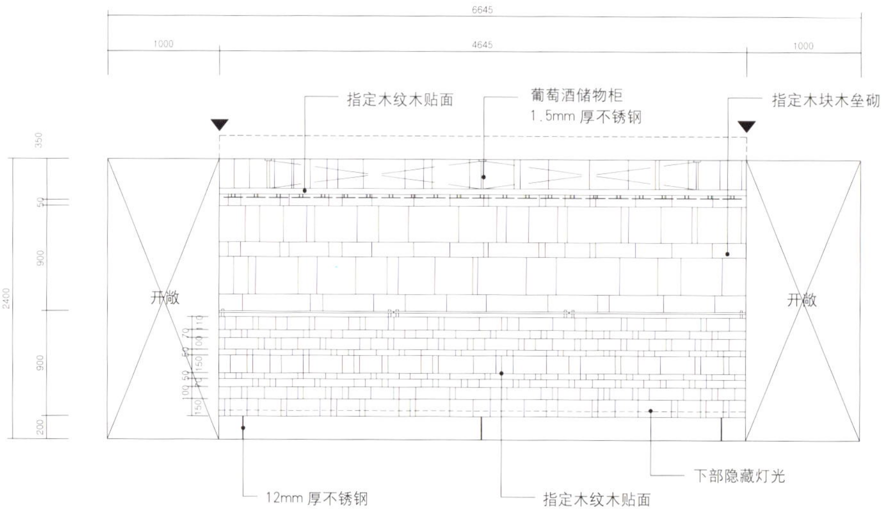

立面图 A

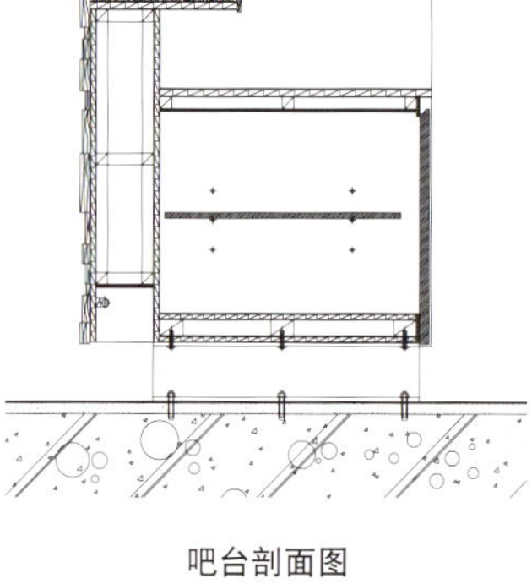

吧台剖面图

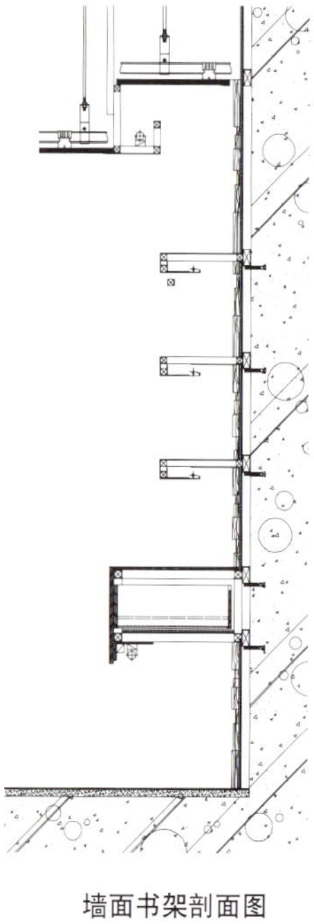

墙面书架剖面图

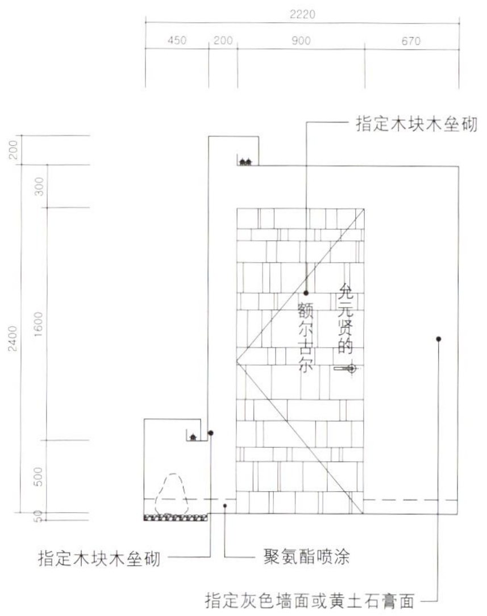

入口立面图

■ 单间一号

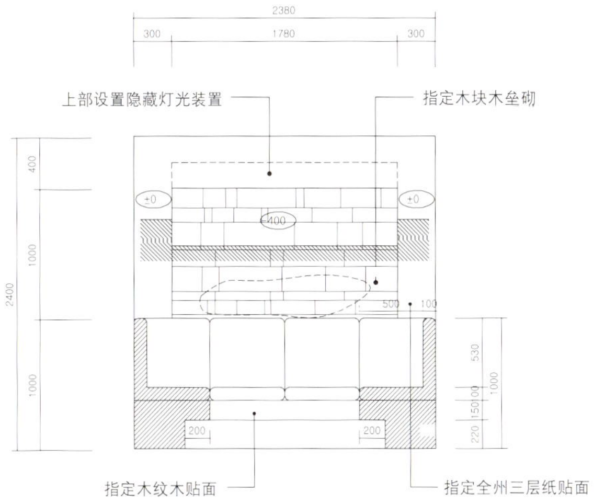

立面图 A

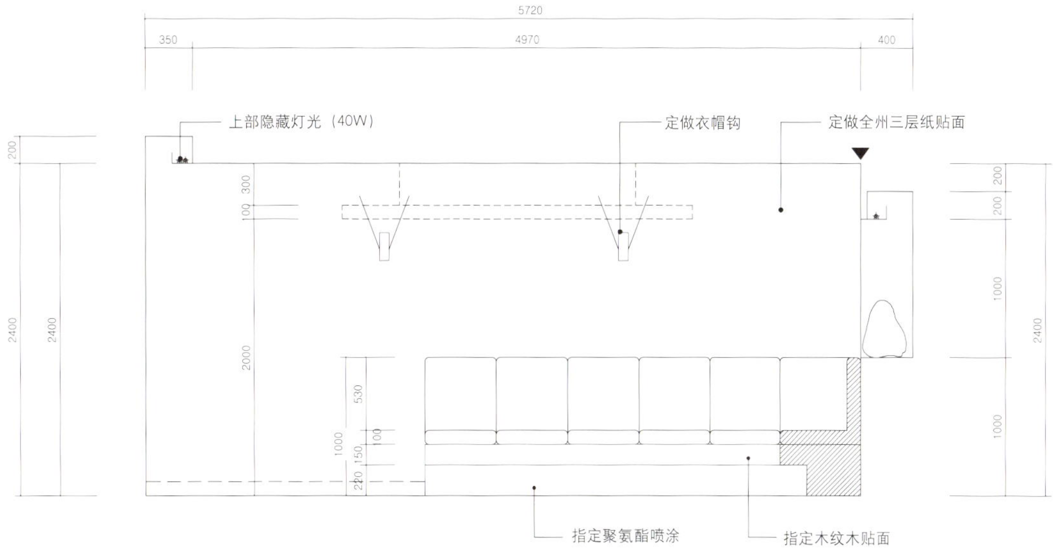

立面图 B

■ 卫生间

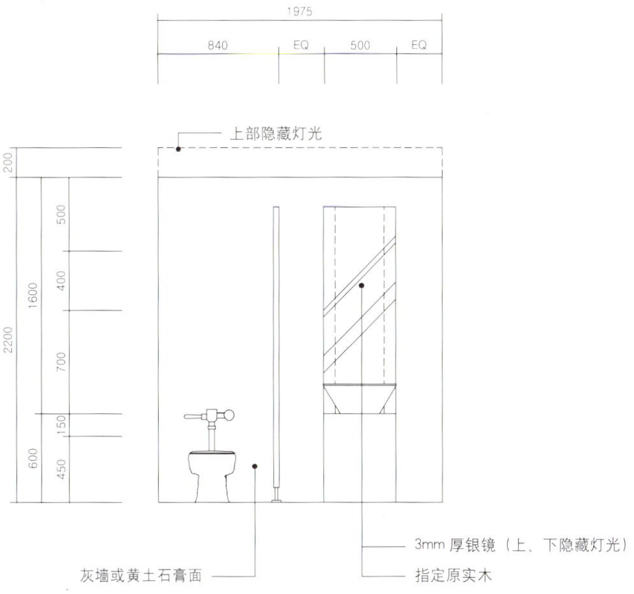

立面图

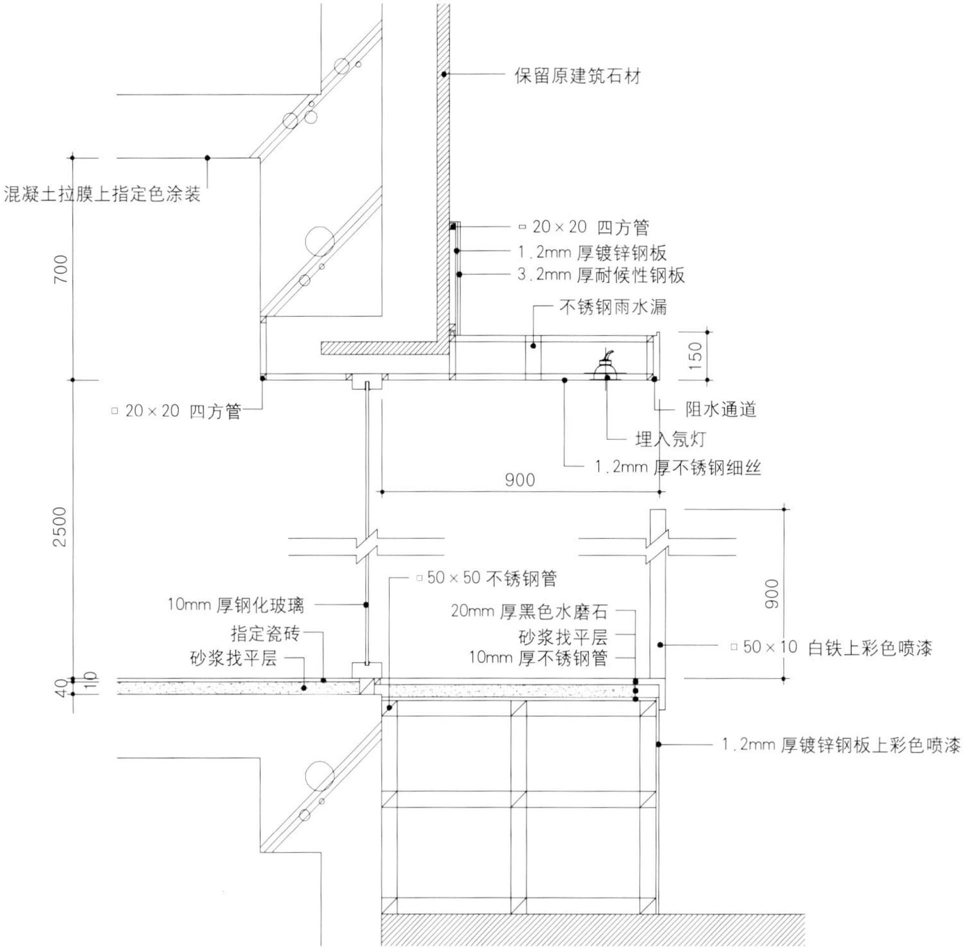

正立面细部图 A（单位：mm）

乐聚缘餐厅是集葡萄酒商店、酒吧以及葡萄酒研究多种功能于一体的场所。业主提供的条件有三个房间、一个较大的葡萄酒销售间、一处零售葡萄酒的大厅及附属空间。如何通过空间布局对“葡萄酒”这一主题的购买、品尝、研究的行为进行解析是摆在设计师面前的重要问题。

葡萄酒追求多种要素中的均衡和协调。如果能在室内空间布局中体现这一特点将会是非常诱人的。因此，该餐厅采用了“清末上海”的模拟真实空间。

18 世纪末至 19 世纪初的上海是近代和现代文化的过渡期，也是东方与西方两大文化的碰撞结合点。固有的文化掺入了异质性的要素，通过否定、异动等独特的文化喧噪，形成了那个时期上海特有的文化现象。模拟这种空间就是要表现葡萄酒的均衡性、协调性和异动性。三个房间分别为不同的葡萄酒爱好者放置不同级别的葡萄酒。

第一间房子是为没有品尝过葡萄酒者设计的，在这里人们可以学到葡萄酒的基础知识。因此，此房子的大厅非常开放，谁都可以进来出去，就像农村的公共空间一样。第二个房间是用树皮做的外部包装，室内红色为房间主色调。这里能够品尝到葡萄酒的口感和香味，这是享受葡萄酒的阶段。第三个房间孤立地隐藏在售酒商店的后面，是酒类专家的专用空间。虽然，每个房间开放的程度不同，被赋予的形态不同，拥有的功能不同，但是它们在一起组成了一个现实的空间，这就是一个由多种要素混合而成的“清末上海”……

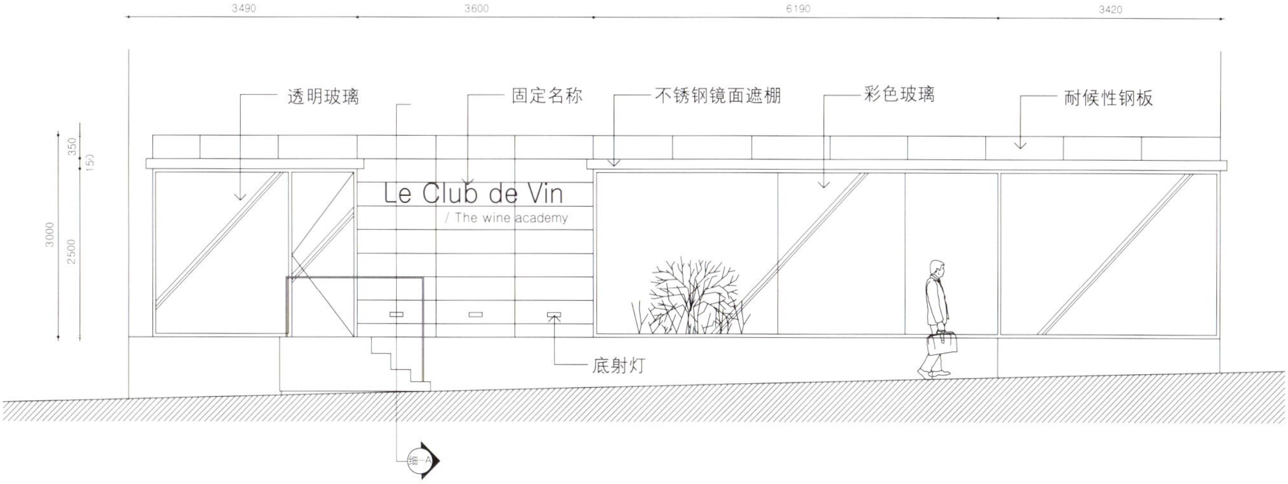

正立面图

设计：（株）李蒙吉加设计公司 金主元 (02) 523–0157 | **施工设计**：（株）李蒙吉加设计公司 希胡 李正兰 | **施工**：（株）李蒙吉加设计公司 高准成 金仓吉 | **地址**：首尔市江南区清潭2洞3–21 留南大厦一层 | **面积**：173m² | **装修**：外部／耐候性钢板 地板／瓷砖、实木地板 墙面／树脂涂装、实木板、收尾后聚氨酯喷涂 顶棚／树脂喷涂、实木板

Design: IMONGGIGA · Kim, Joo won | **Execution Design:** IMONGGIGA · Seo Ho / Lee, Jeong ran | **Construction:** IMONGGIGA · Go, Jun seong / Kim, Chang gil | **Built Area:** 173m² | **Finish:** Exterior - Corten Steel Plate Interior - Floor / Tile, Wood Panel Wall / V.P Paint, Wood Panel, Urethan Paint over Plaster Ceiling / V.P Paint, Wood Panel

■ 结算台

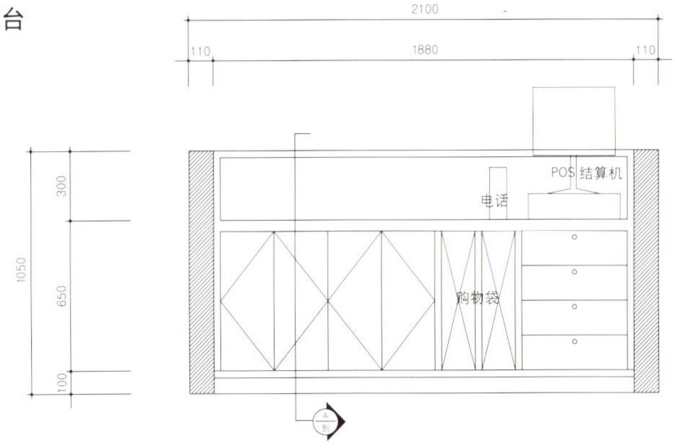

立面图

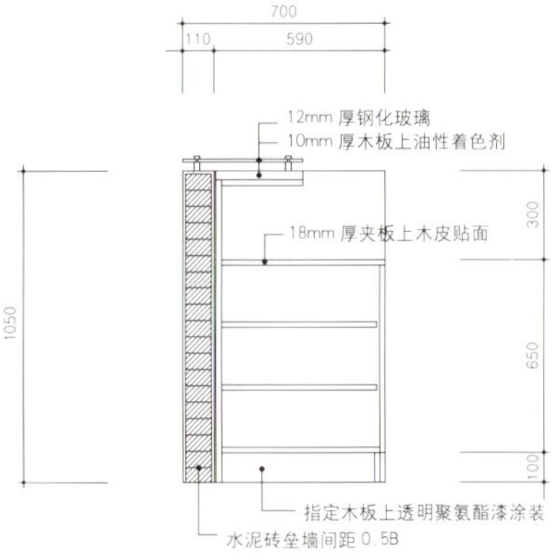

剖面图 A

■ 大厅

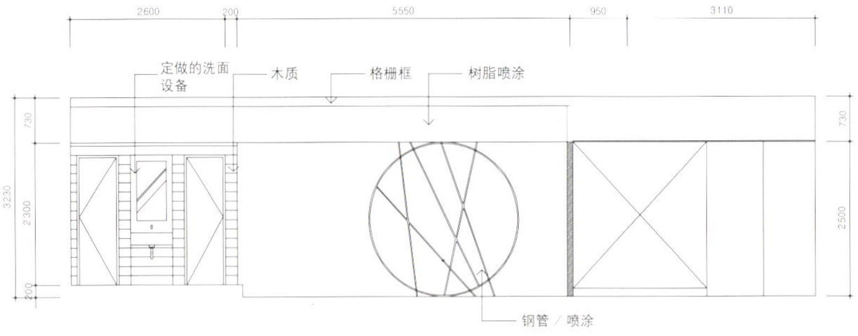

立面图 A

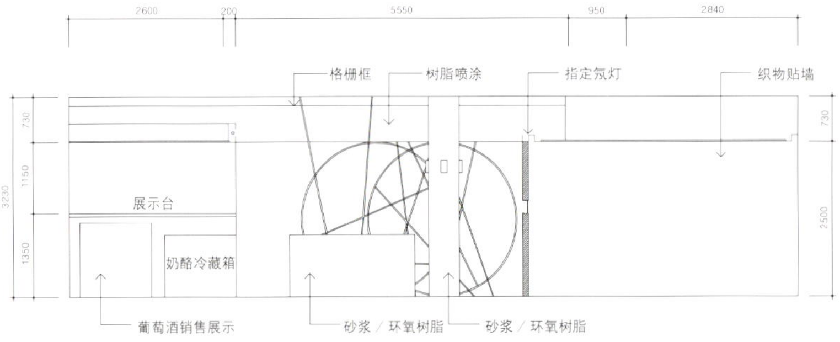

立面图 B

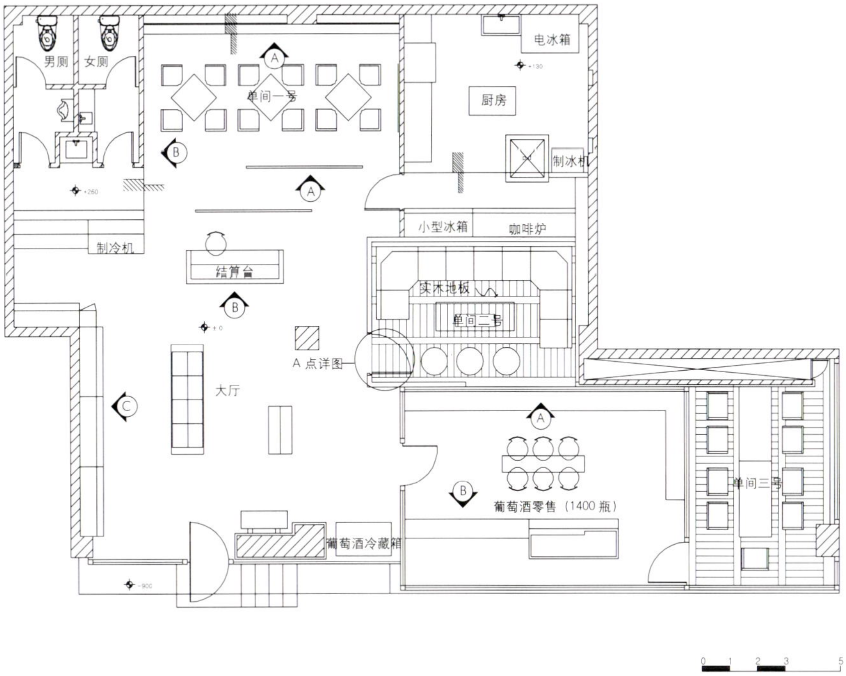

平面图

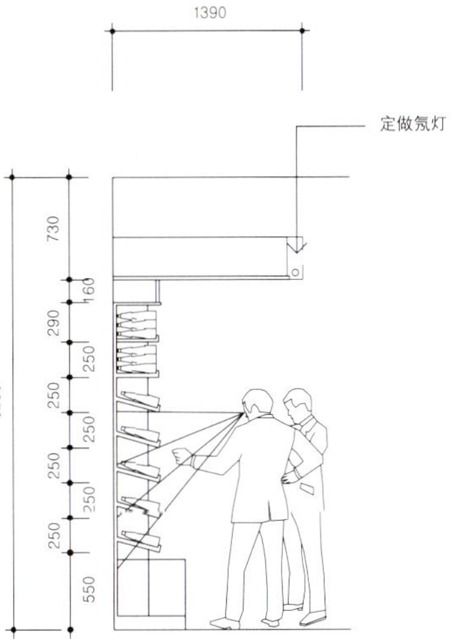

剖面图 A

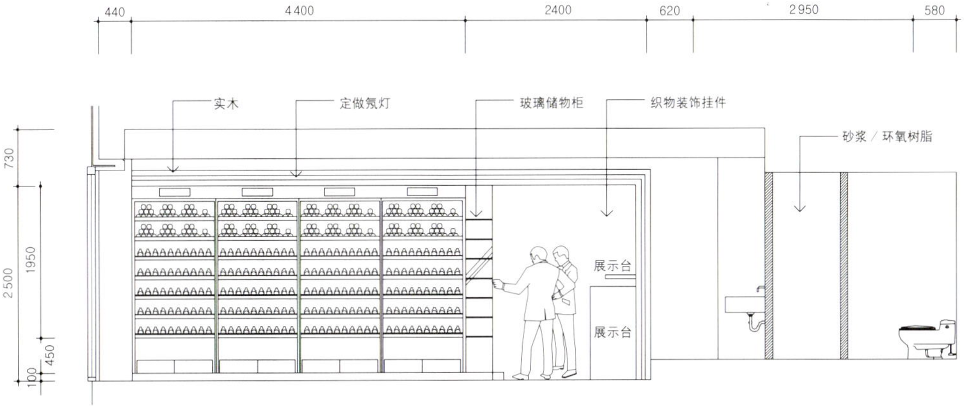

立面图 C

The complex place is combined of wine shop, bar and academic facility. In detail, three function rooms, a wine-seller room in pretty large size, retail space for wine selling in the notion of 'Hall' and the others are laid out in the place. The main concept is addressed to the theme of 'wine' and the design solutions of combining wine with buying drinking and learning it.

wine pursues the balance and harmony in the chaos of various elements. This characteristic of wine is desired to represented in the place. Finally, the space-time of Shanghai in Chinese Qing Dynasty is adopted. The city from the late 19th century to the early 20th century is the point of contacting modernism with pre-modernism, and Chinese tradition with Japanese and Western cultures.

The three rooms are classified for different level of wine enthusiasts. The first room is designed for the beginner to achieve basic knowledge on wine, where is closed and opened to the hall, such as public place to access easily. The second red room, surronding by bark, expresses the taste and flavor of wine, that is, the level of enjoyment. The third room is isolated behide the wine-seller room, the cldsad world of experts. Each room is designed in different formation, the grade of openness-closing and a role, and is combined of a unity of space-time at the same time. Just like the city of shanghai in the late period of Qing Dynasty with mixing various elements...

■ 单间一号

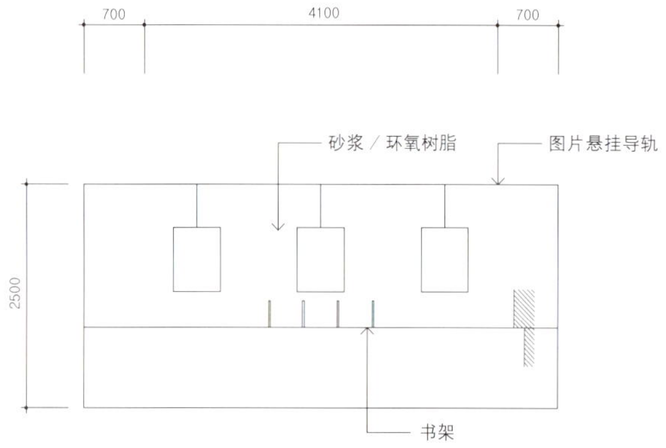

立面图 A

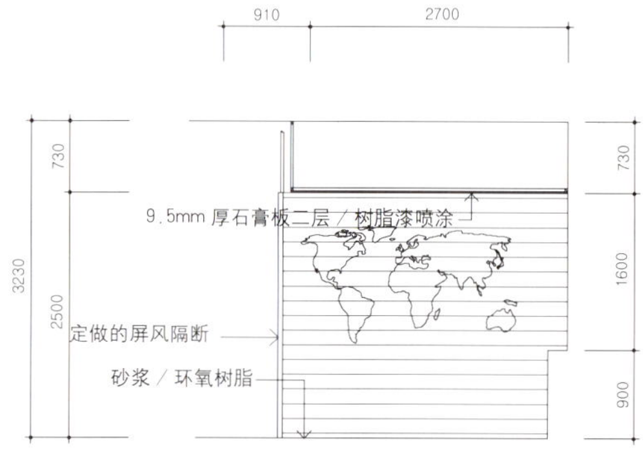

立面图 B

■ 单间二号

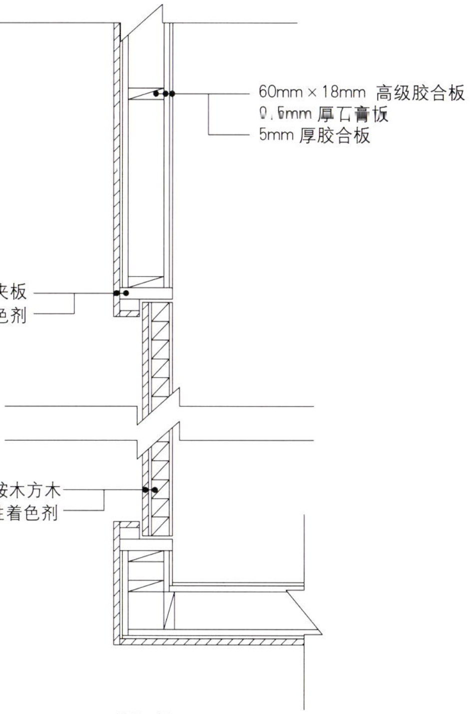

门细部

■ 葡萄酒销售

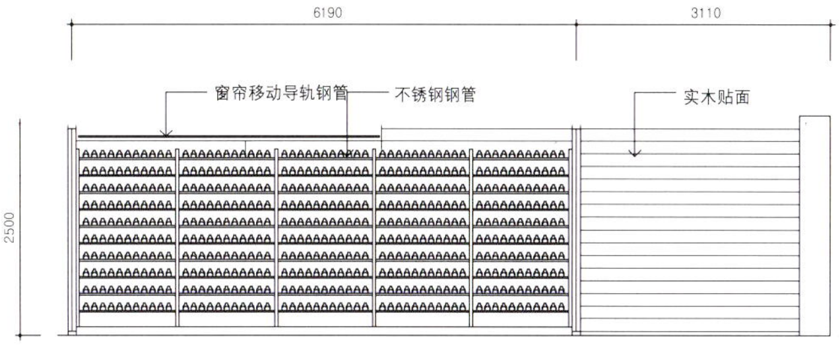

立面图 A

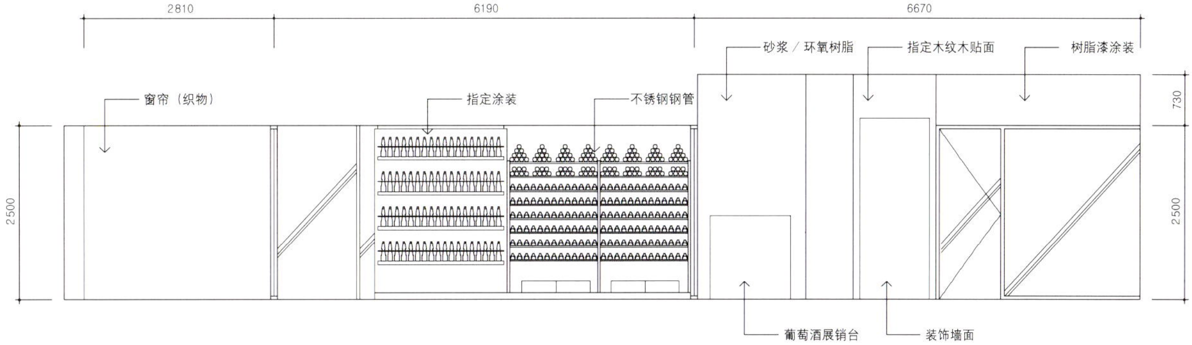

立面图 B

■ 结算台—外卖口

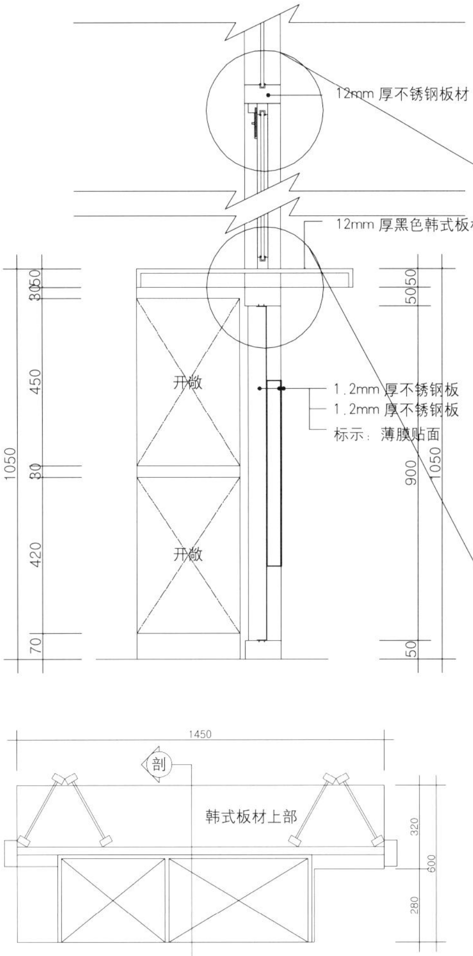

结算台与外卖口—局部平面图

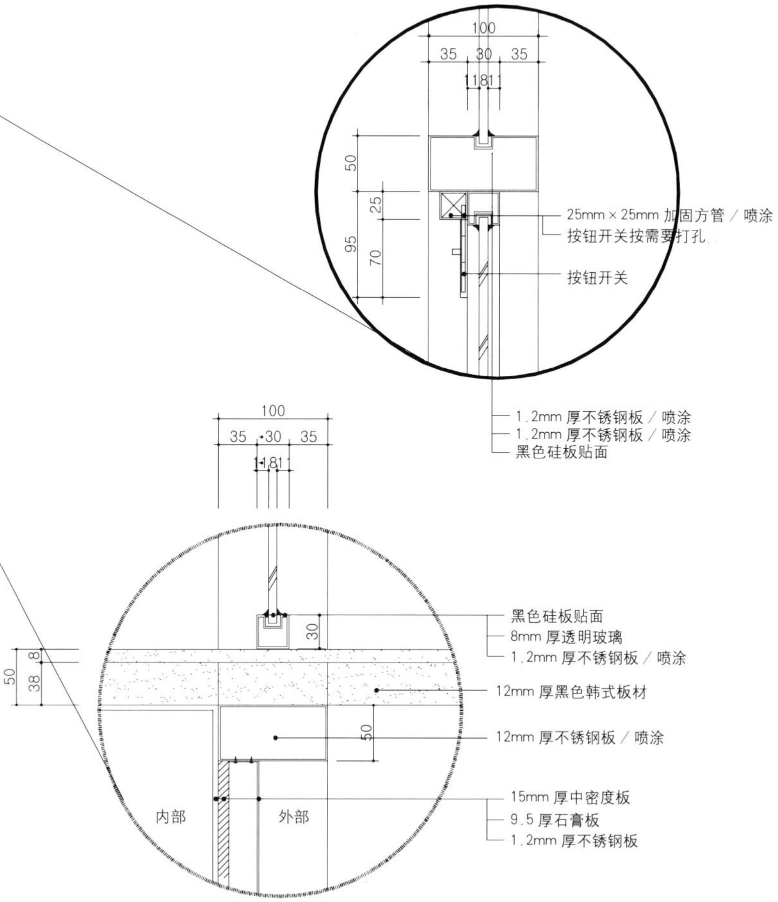

结算台与外卖口—剖面和细部图

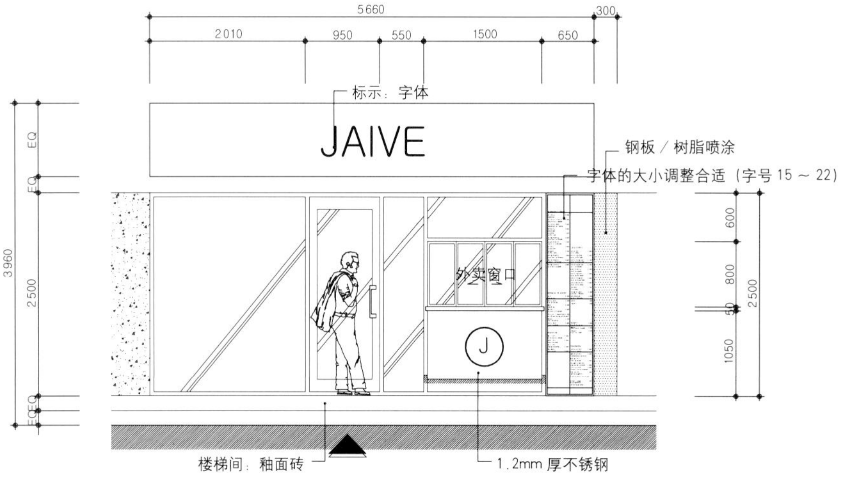

正立面图

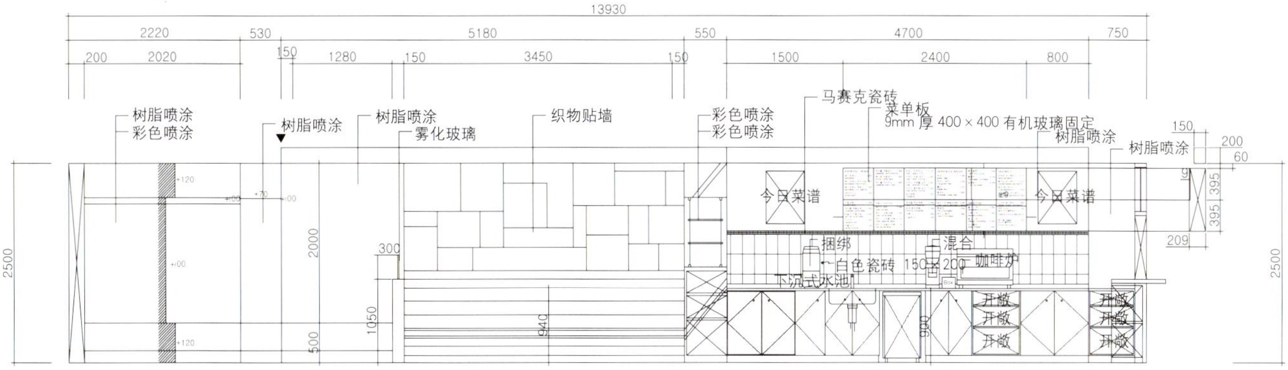

大厅立面图 A（单位：mm）

设计：设计城市公司 朴之龙 (02) 517–3370 | **施工**：设计城市公司 韩民 朴之英 | **地址**：蔚山市南区无举洞 618–10 番地 | **面积**：92.93m² | **装修**：地面 / 釉面砖、陶瓷锦砖 墙壁 / 涂装、彩色玻璃、瓷砖、不锈钢细线、纤维 顶棚 / 涂装 | **合作公司**：地面 | 吴成瓷砖 (02–424–4960)、金属 / 亚洲金属 (011–282–2196)、木工 / 格林公司 (02–379–4028)、家具 / 凤凰社 (02–3452–4946)、详细家具咨询电话 (02–542–0244) 签字 / 循设计公司 (02–566–4960) 设备 / 大宇建筑设备 (02–977–0404)

Design: Design City · Jeffrey Park | Construction: Design City · Han Min / Park Jin young | **Built Area:** 92.93m² | **Finish:** Floor / Polishing Tile, Mosaic Tile Wall / Paint, Color Glass, App.Tile, S'st Hairline, App.Fabric Ceiling / Paint

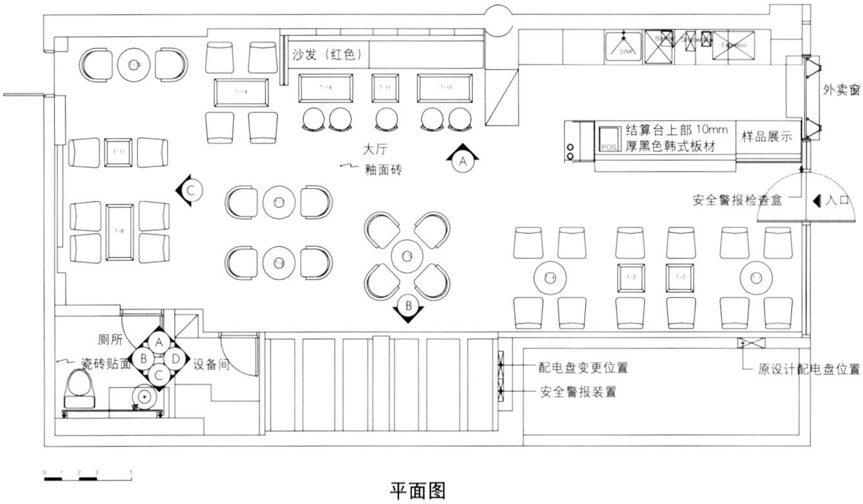

平面图

■ 结算台

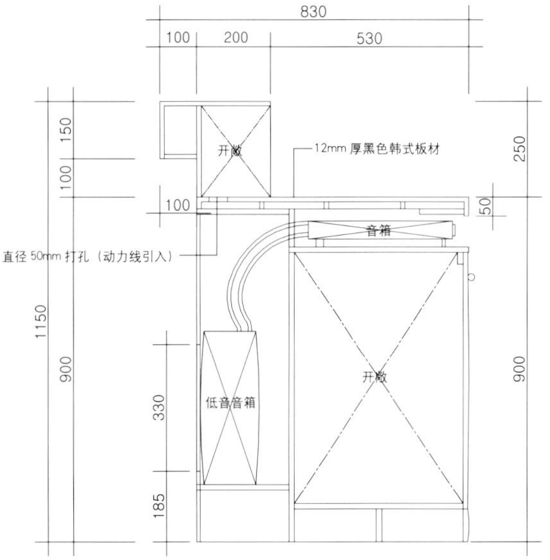

剖面图

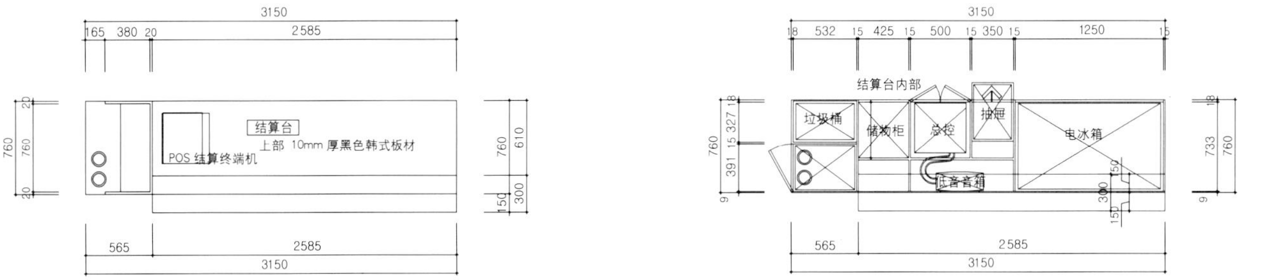

顶视图与细部图

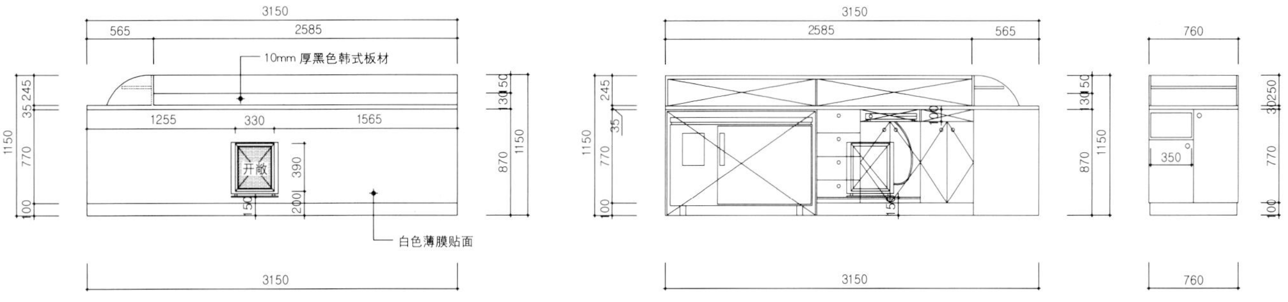

前视图、后视图、侧视图

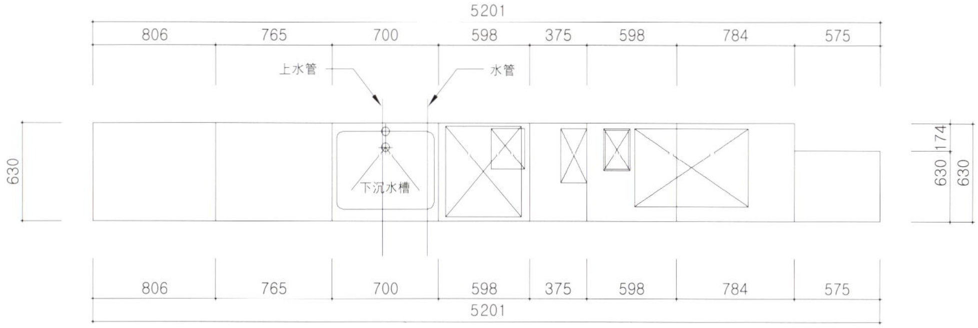
5201
806
765
700
598
375
598
784
575
上水管
水管
下沉水槽
630
174
630
630

顶视图

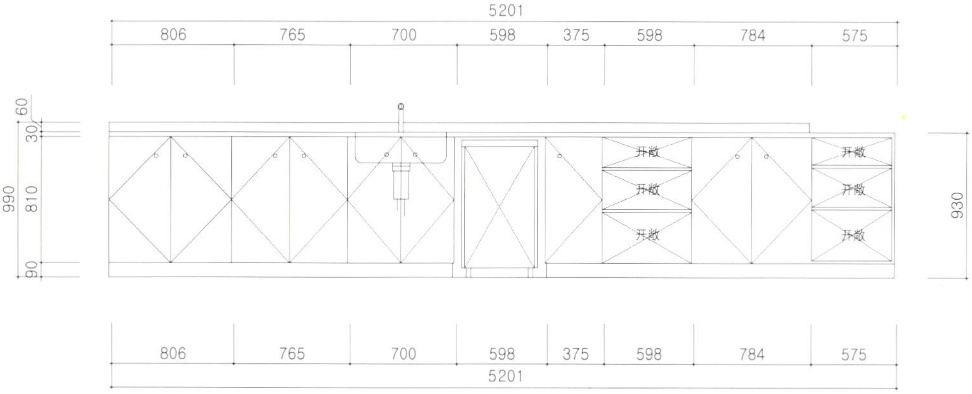
5201
806
765
700
598
375
598
784
575
60
30
990
810
90
930
开敞
开敞
开敞
开敞
开敞
开敞

前视图

■ 大厅

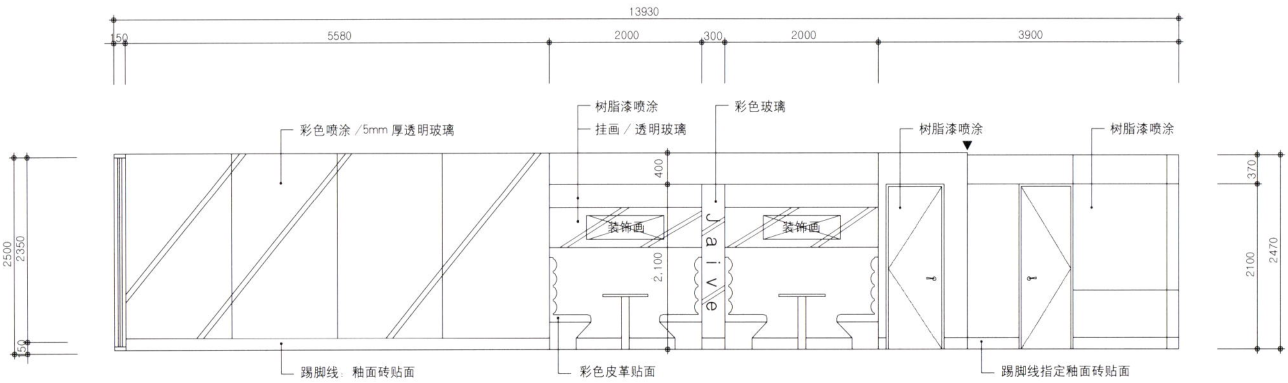

立面图 B

翟氏咖啡屋是一家咖啡专卖店，它是伴随着现代社会流行的快餐文化而出现的，喝咖啡也是饭后消遣的一种方式。咖啡是什么呢？咖啡不是指单纯的换换口感的饭后消遣，而是要达到口感和气氛的统一。对于咖啡文化，具有专业化的咖啡店才算是跟上时代潮流的。咖啡的主要消费对象是青少年和中年人，如果考察一下他们的生活方式，为他们提供快捷、简便、高级的专业咖啡，是咖啡店或者称之为咖啡屋立于不败之地的根本保证。现在专业的咖啡外卖竞争非常激烈，因此，应该为此创造一种更加有效的空间环境，例如：提供更便宜、质量更好的咖啡原料，理顺咖啡的销售渠道，提供周到的服务等。

“外卖”店面的强烈的颜色对比（黑色和白色）比其他咖啡屋具有压倒性的视觉效果。咖啡店的醒目标志可以让顾客很容易地发现它，挂在室外的墙上的菜单板能够让顾客很方便地选择他们喜欢的咖啡。

“销售－结算台”：站在结算终端机前可以看到菜单墙上的各种各样的咖啡价格信息。

“顾客服务处”：位于室内，地面和家具采用了强烈对比的翟氏咖啡色（黑与红），墙面和顶棚则混用了拿铁咖啡的浅棕色，这样的颜色搭配显得更丰富、更安全也更柔和，是客人们理想的休息室。

摆放座席时选择的沙发和椅子是固定的，餐桌则提供了几种组合，一张桌子形成的独立式或几张桌子拼起来形成的团体式。照度调节器可以调整室内光线，以满足不同的氛围要求。白天和夜晚的照度是不同的，白天选择明亮轻快，夜晚选择典雅安逸。

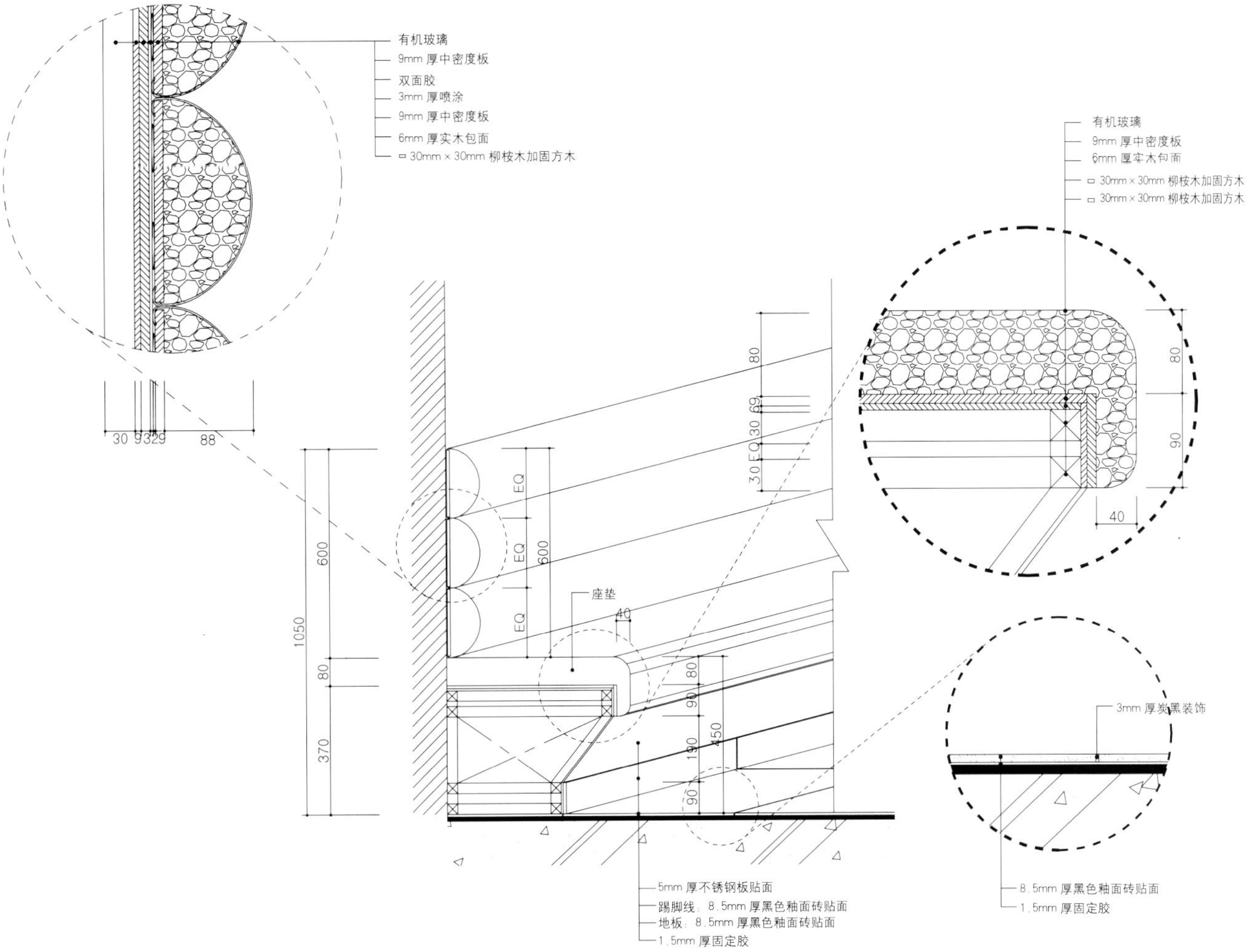

固定座椅—剖面和细部图

Coffee shop chain 'Jaive Esspresso' is a dessert born from modern 'fast' culture. Coffee is not just a simple dessert masking the aftertaste anymore; since it is even said that you can say you drank one only when the taste is companied by atmosphere, a professional coffee shop is in definite need in this time.

Considering the main customers ranging from teenager to the middle-aged, a shop has to be designed to serve a high quality coffee quickly and conveniently.

A take-out coffee shop market is in the middle of a fiercely competition. Thus Jaive Esspresso aimed to present an effective space providing a successive sale and smooth service, and to planting the image of a shop serving a high quality coffee in lower price.

Exterior / Take-out

Combination of the intense colors (black and white) outside is the first to get this shop to be noticed prior to the others. An awning enables anyone to take the space as a take-out coffee shop, and an exterior menu board was installed for accessibility from outside.

Sell / Counter

Counter : The menu board in front is the most noticeable installation.

Menu Board : We approached with a functional design for effective recognition of the diverse menu, applying emboss finish and color arrangement.

Customer Service Area

Located on the far inside. The floor and furniture in Jaive's point colors (black and red) show a striking contrast, and the wall and ceiling is in secondary light brown of cafe latte, creating a customer lounge with a diversity, comfort and softness. Fixed sofas and chairs enable the flexibility in switching group and individual seats. Organic atmosphere comes from light intensity alteration. By adjusting the dimmer, lighting for day and night changes; bright, cheerful in the day, and sentimental and cozy at night.

ESPRESSO COFFEE DRINKS & ICED WATER

■ 大厅

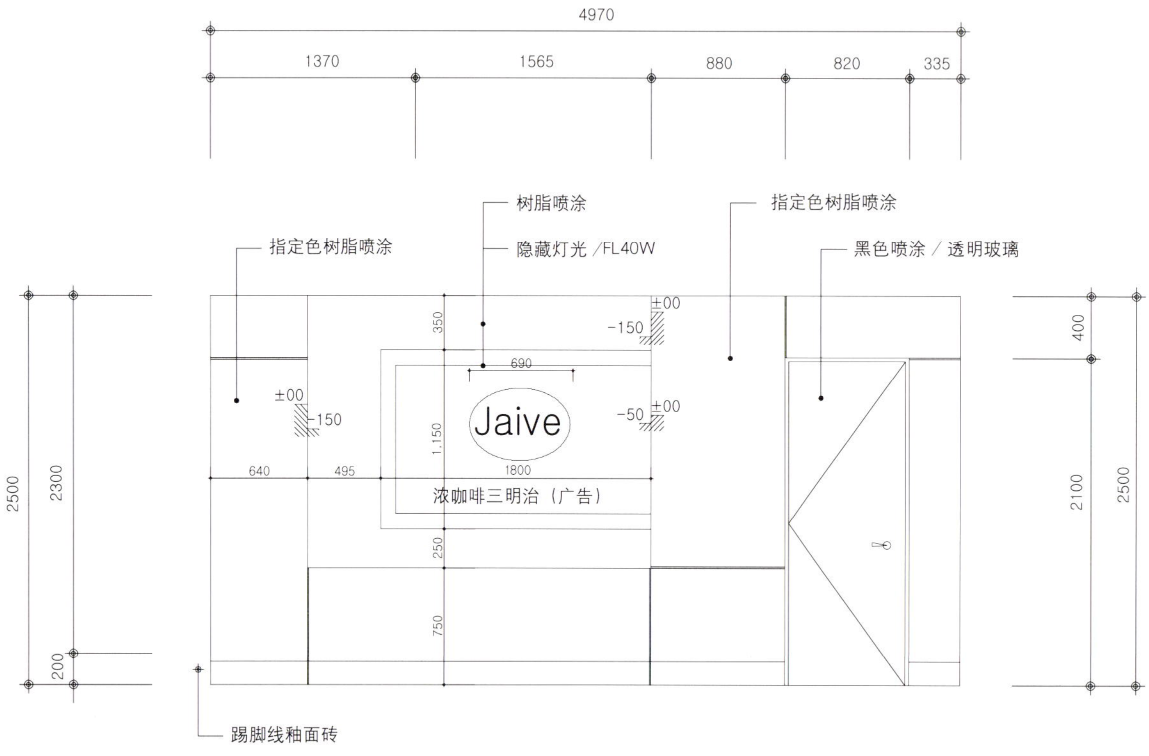
4970
1370
1565
880
820
335
树脂喷涂
隐藏灯光／FL40W
指定色树脂喷涂
指定色树脂喷涂
黑色喷涂／透明玻璃
±00
-150
350
690
Jaive
-50
±00
1,150
640
495
1800
浓咖啡三明治（广告）
250
750
2500
2300
200
400
2100
2500
踢脚线釉面砖

立面图 C

■ 卫生间

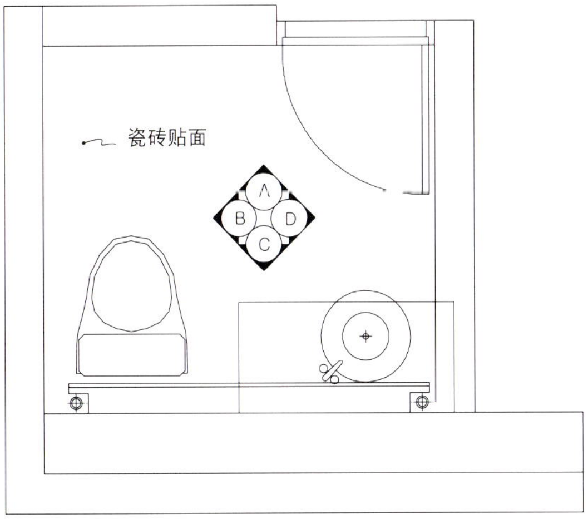

局部平面图

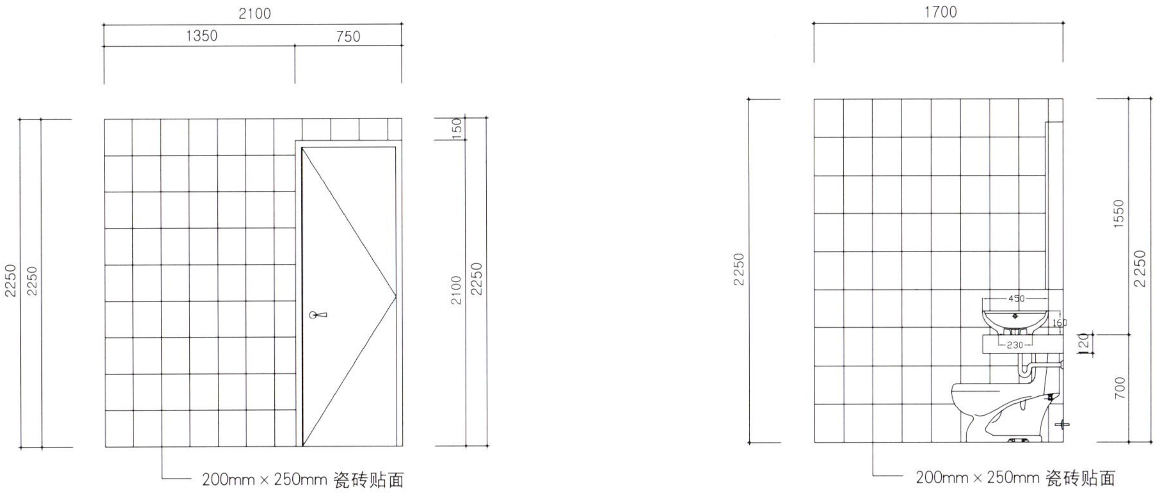

立面图 A、B

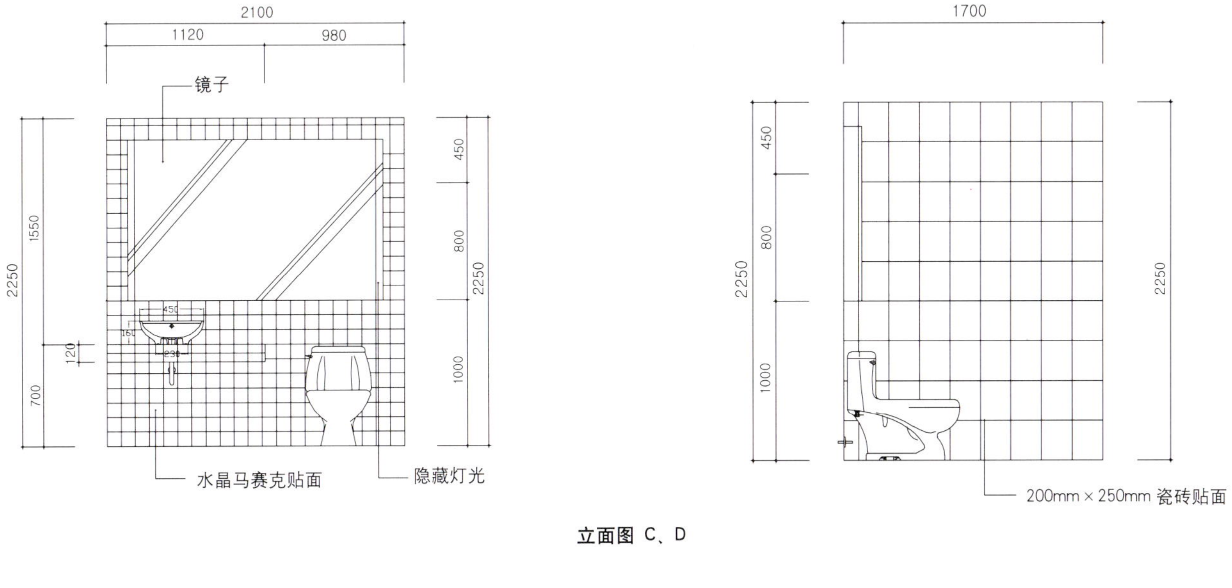

立面图 C、D

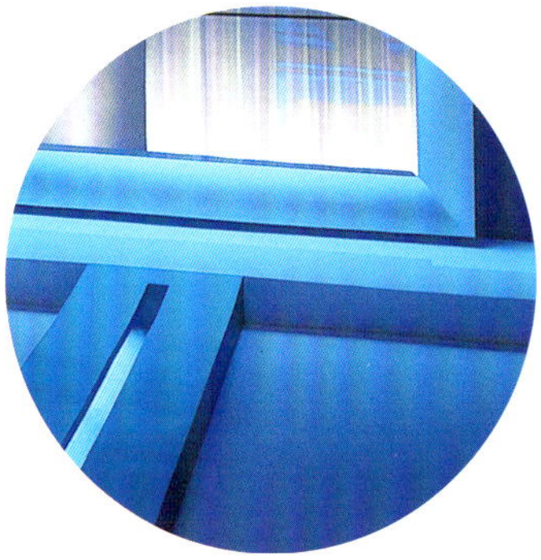

艾丽菏歌区
TTL 歌区釜山西面店
明星之塔

罗拉会员店
浦项清静茶礼园

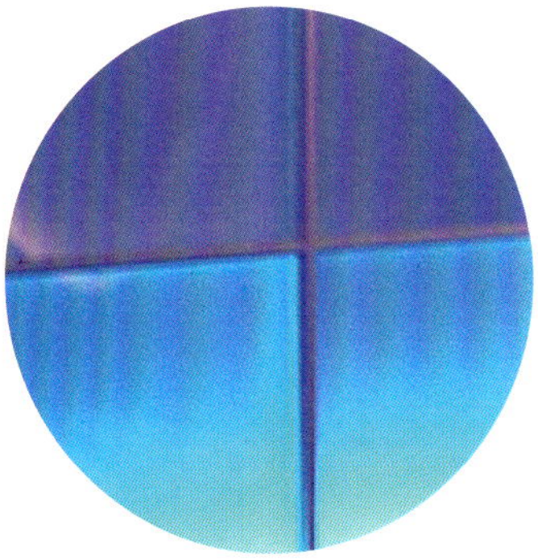

包嗒饰品店
极乐歌区大学路分店

最新室内细部设计实例集Ⅵ

娱乐场所

Recreation

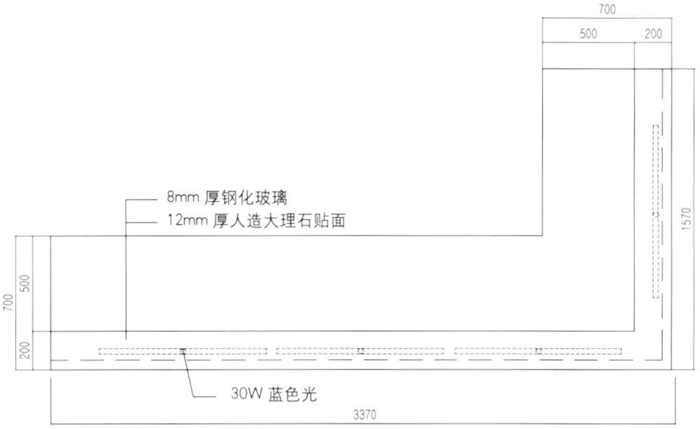

服务台—顶视图

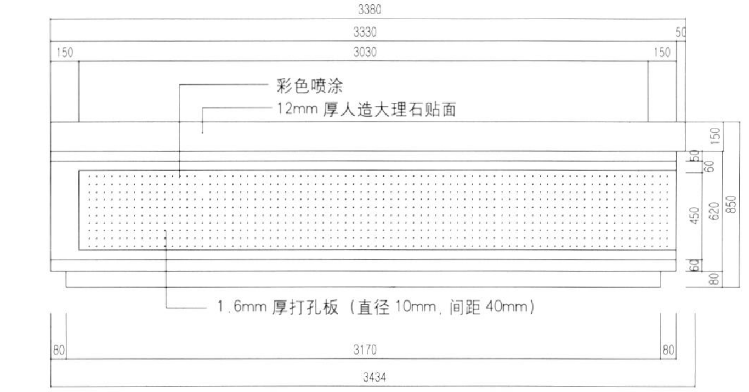

正视图

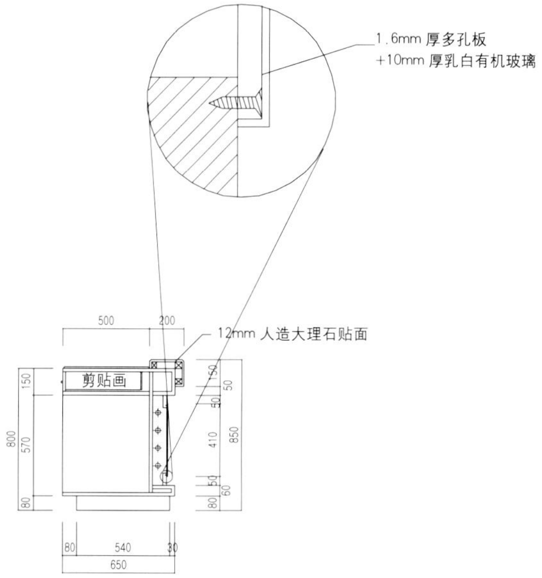

剖面和细部图

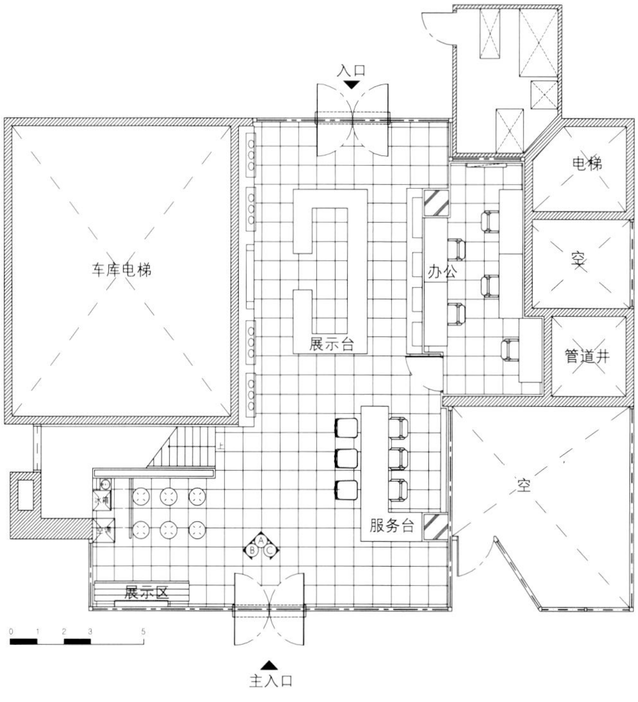

一层平面图

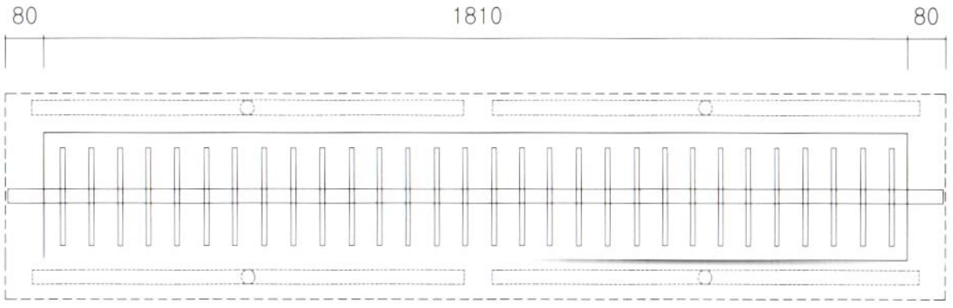

服务台照明—顶视图

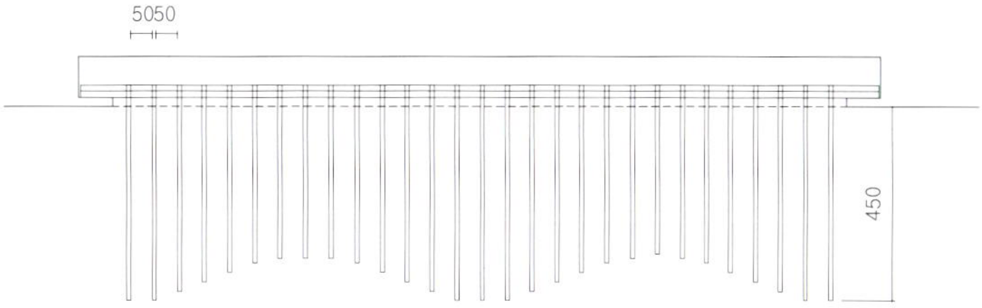

正视图

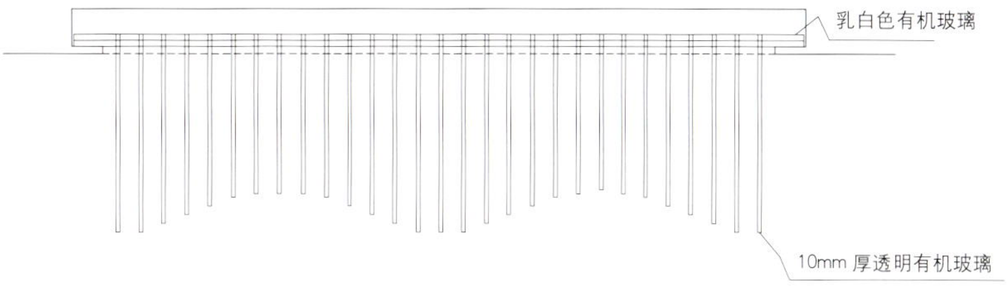

正视图

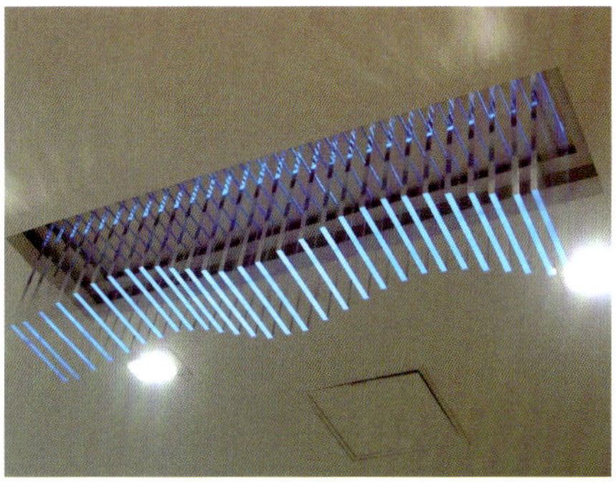

设计：麦赛设计公司　金太（02）546-7797 | **施工**：麦赛设计公司　朴尚根 | **地址**：首尔市钟路区东崇洞 1-33 | **面积**：281m² | **装修**：地面／瓷砖、玻璃墙面／彩色玻璃、乳胶漆、玻璃瓷砖、钢材、B/M STS 顶棚／树脂喷涂、钢材、有机玻璃

Design: Design Messe · Kim, Tae | **Construction:** Design Messe · Park, Sang gun | **Built Area:** 281m² | **Finish:** Floor / Tile, Glass Wall / Color Glass, Lacquer, Glass Tile, Steel, B/M STS Ceiling / V.P, Steel, Acryl

■ 外部

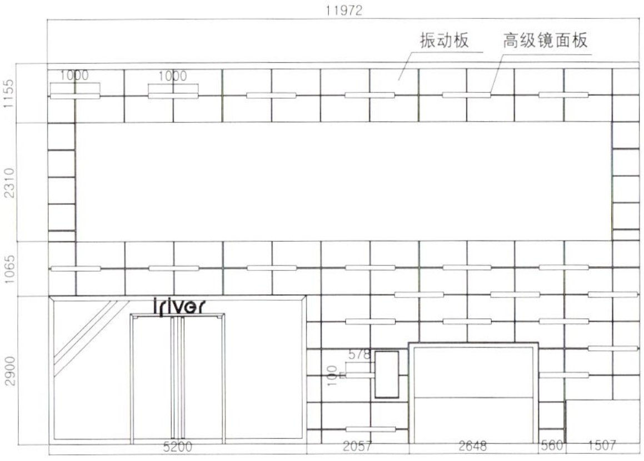

后视图

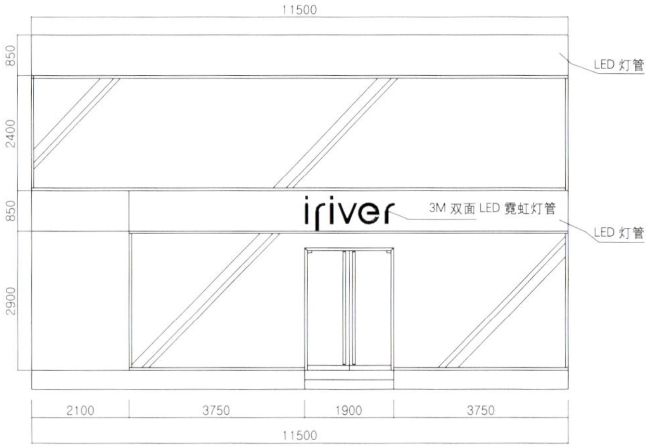

正视图

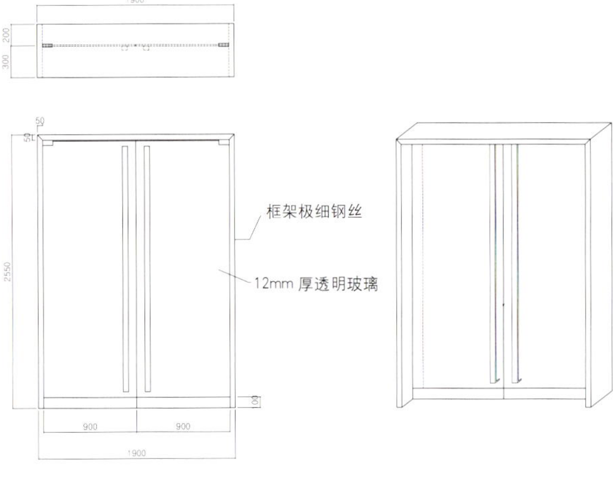

主入口门—顶视图和侧视图

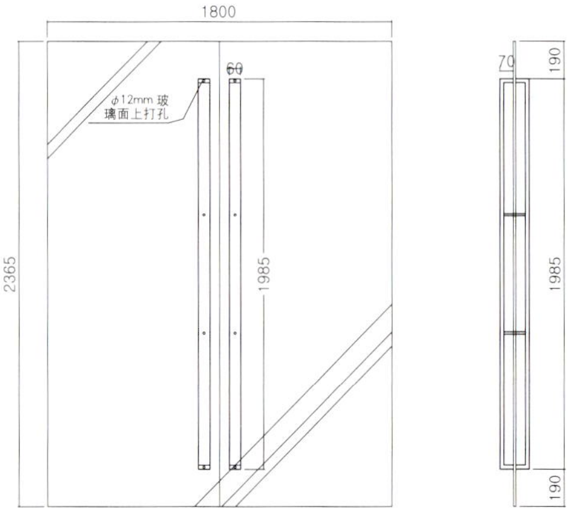

门把手—正视面和侧视图

■ 厅

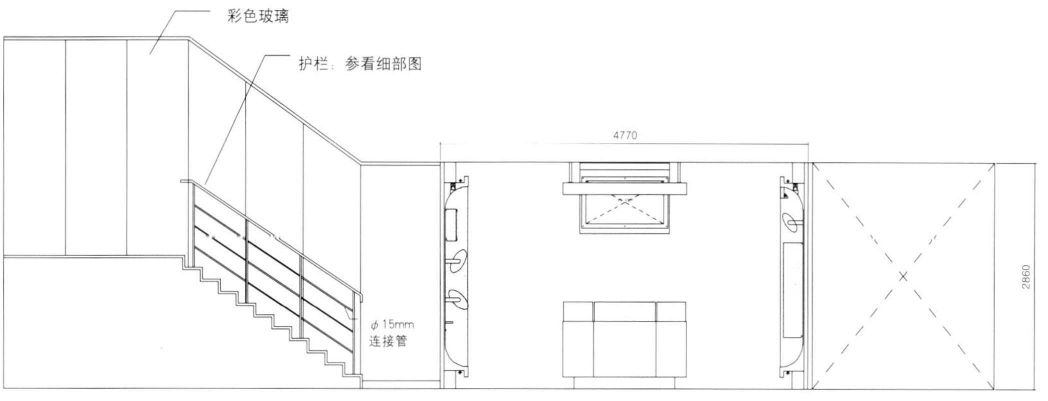

立面图 A

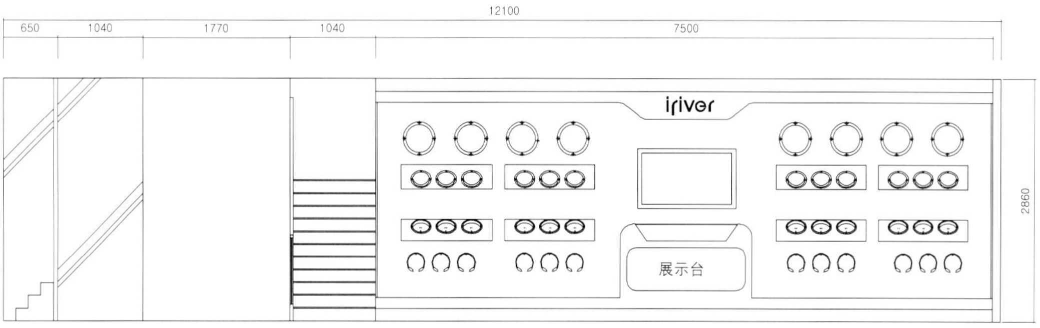

立面图 B

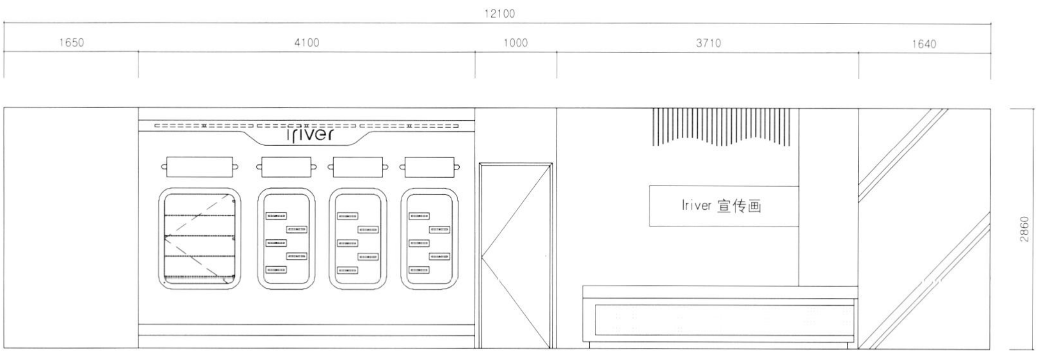

立面图 C

由于项目处在城市中心的美观地区，所以，施工过程中规划部门的要求颇多。而且，从该项目的所处的位置来看，也要求快速施工缩短工期。因此采取了室内外同时开始建设的方法，所以，工程进行得比较困难。工程的重点也放在了LED照明的效果和尺度以及建筑结构加固及防水处理上了。

在这种背景下，完工的大学路艾丽蒿歌区的一层是产品售后服务处，服务台的上面设计了随音乐跳跃的EQ，给人以活泼的感觉。室内布置了欣赏MP3的演示台和MP3机的展示柜台。

建筑的一层整体采用了白色和金属色作为主色调，表达了洁净和律动。二层是为来此进行产品修理的顾客提供的休息室，同时还是为音乐发烧友提供的聚集地，因此，室内的空间处理成了咖啡座的形式，以利于相互交流。以绿色为主基调，彩虹贴面的柱子给人温暖舒适的感觉。

There were difficulties in the construction progress from the early stage, because we had to work on the exterior and interior at the same time. The focus lies on the effective realization of LED lighting, waterproofing, and reinforcement of structure and scale. Being in an urban esthetic district, the project involved numerous obstacles and required a hurried construction and adjustment of the schedule. Iriver Zone has been built with such a story in the background. On the 1st floor, where technical support for Iriver products is provided, the equalizer on a information desk sways up and down rhythmically, bringing liveliness. Further inside placed are the main display cases and demonstrating areas of mp3 players. White and metal tones are mainly used on the 1st floor and make a clean and active space. 2nd floor has a lounge, a cafe-like service area that can also function as a central spot of community meetings. Overall color is green, with shining rainbow pillar maintaining a warm, comforting tone.

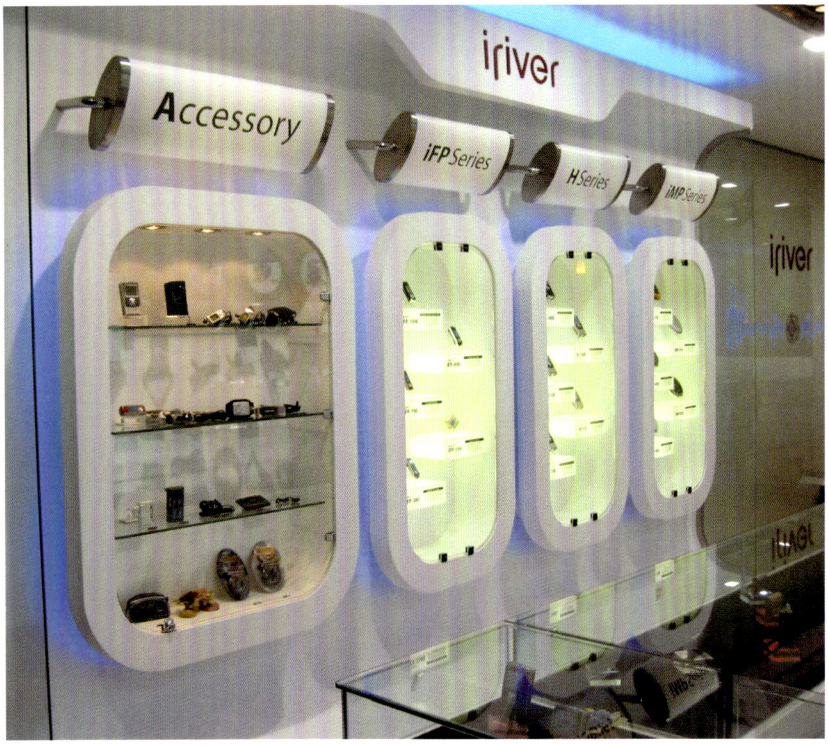

■ 一层公众联系区

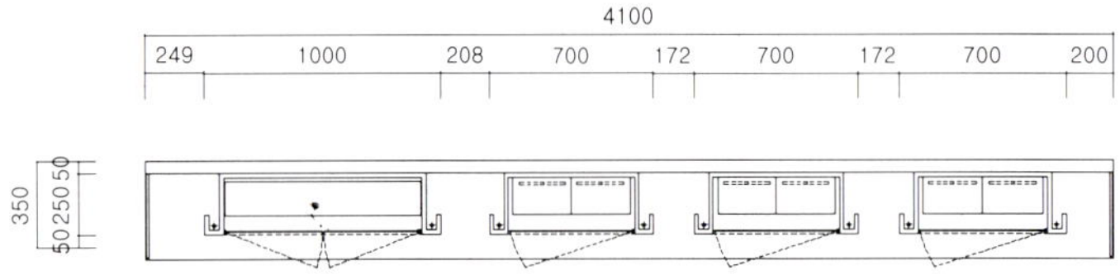

顶视图

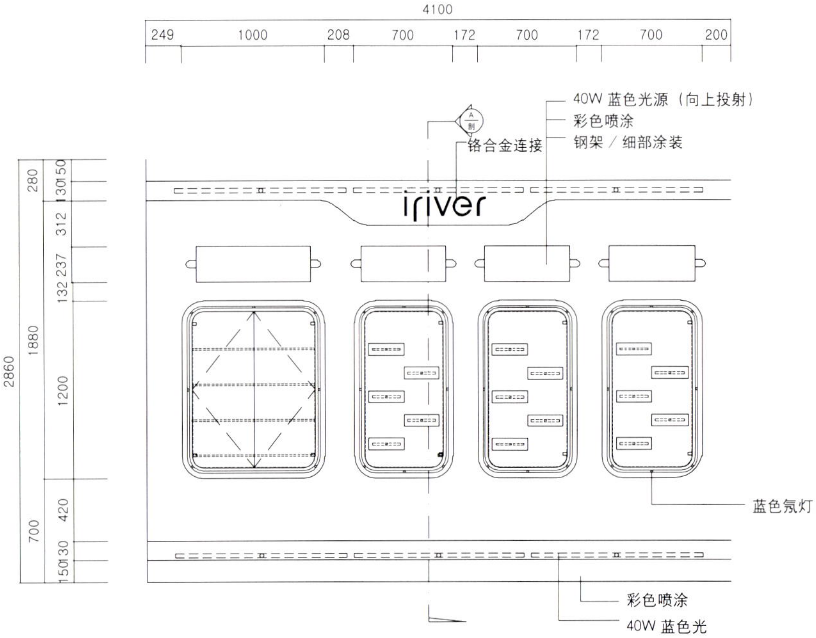

正视图

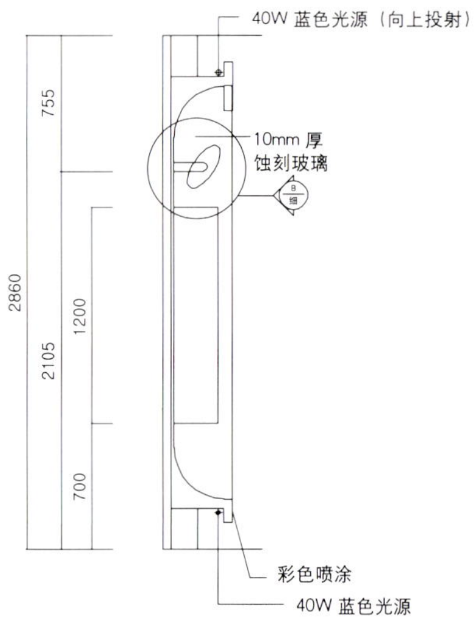

剖面图 A

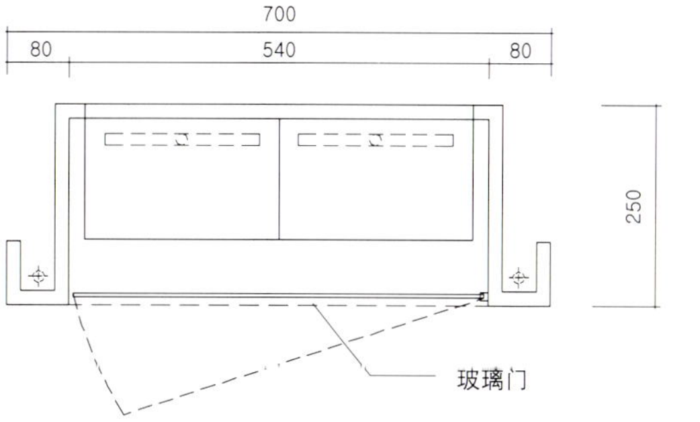

顶视图

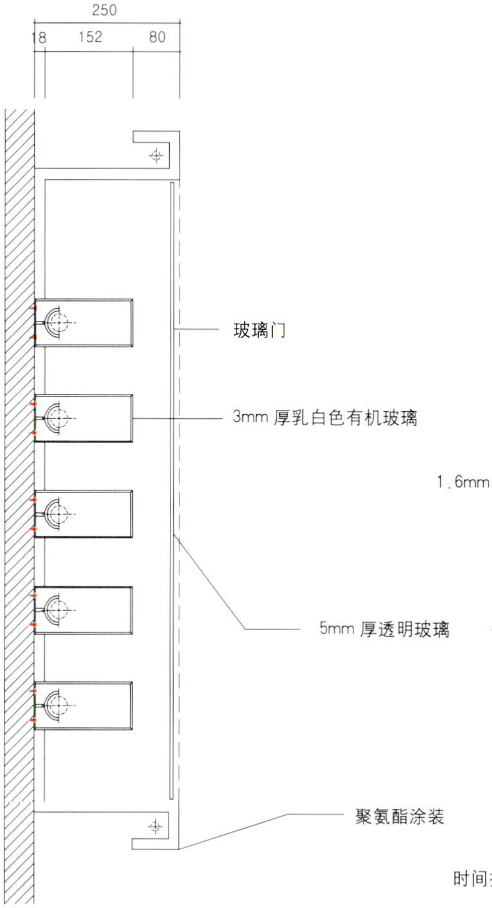

剖面图

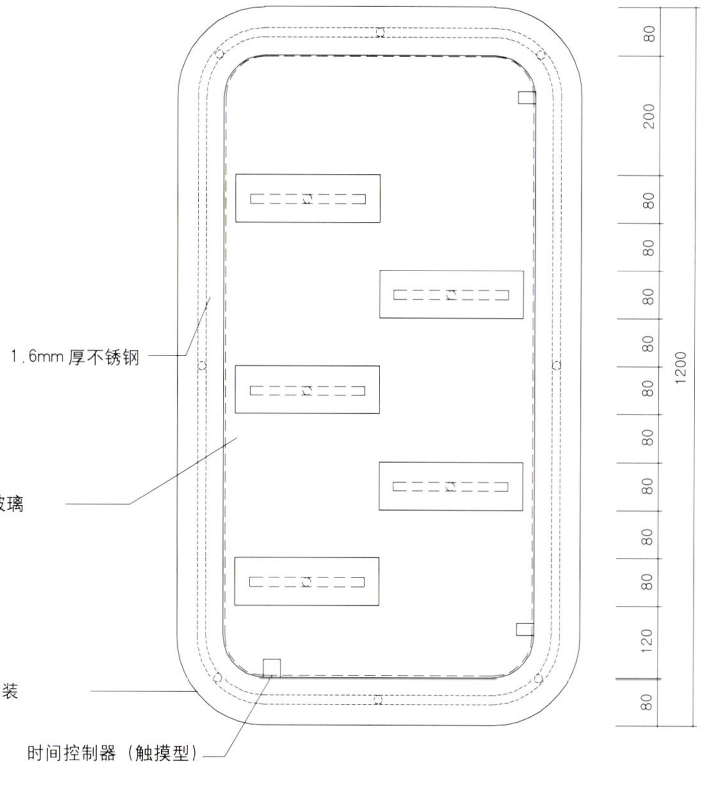

正视图

■ 一层公众联系区

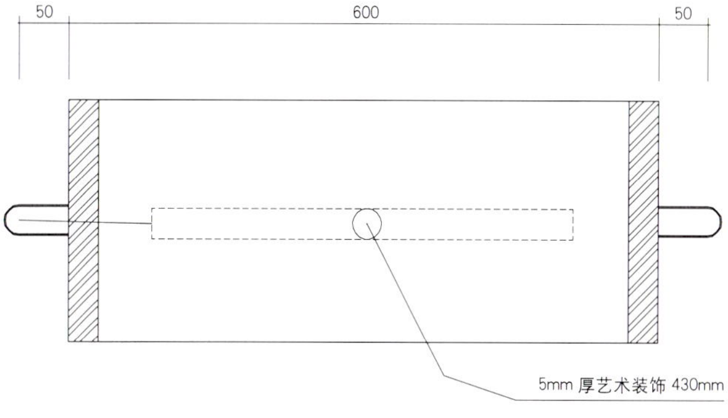

后视图

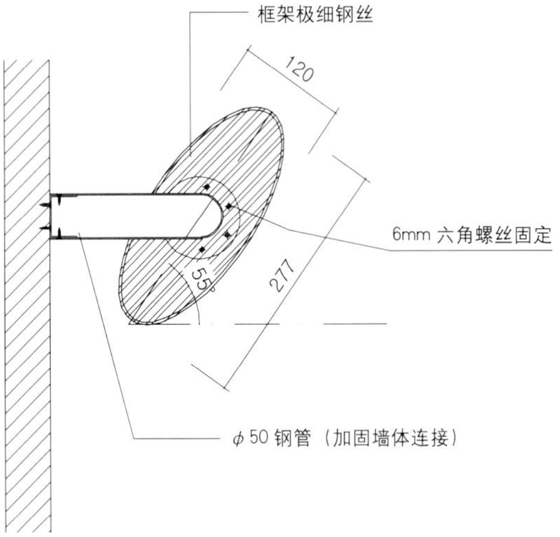

剖面细部图 B

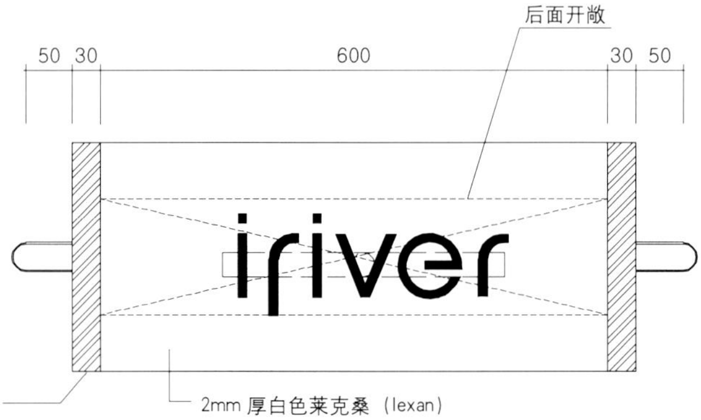

正视图

■ 一层展示区

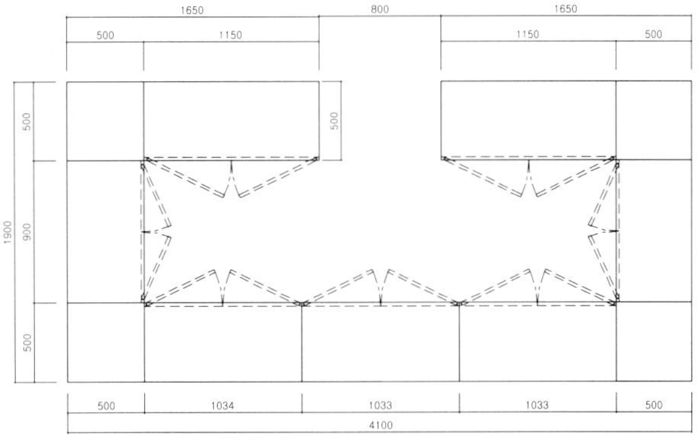

顶视图

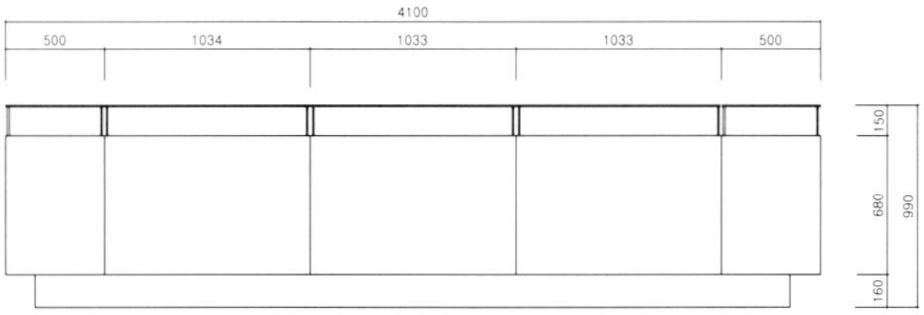

正视图

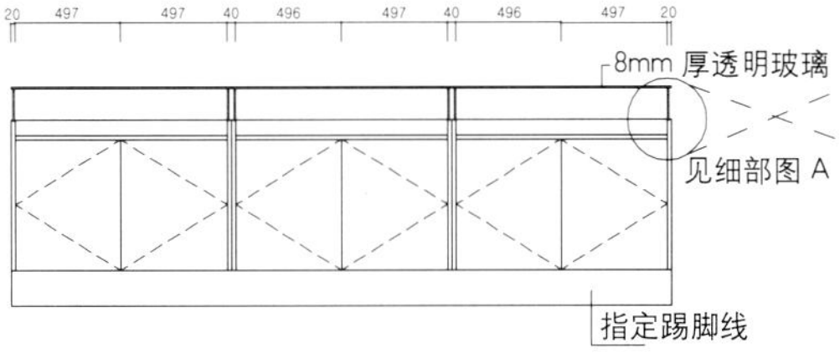

后视图

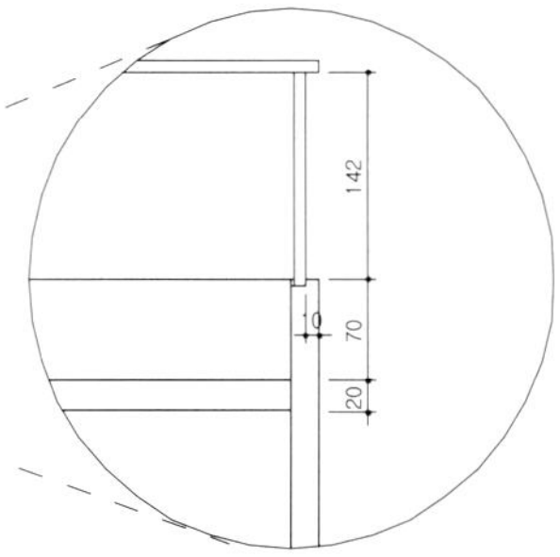

细部图 A

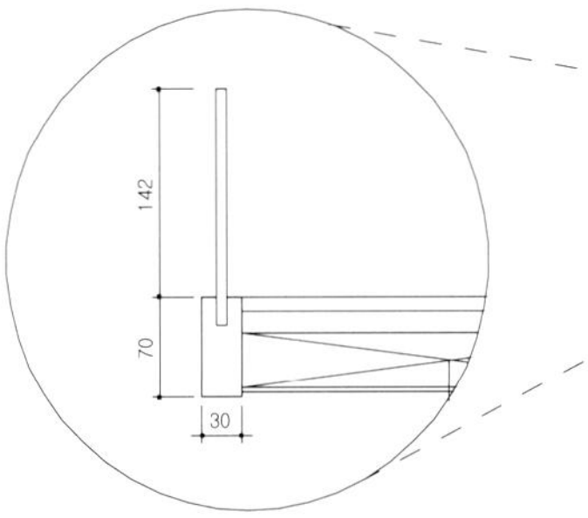

细部图 B

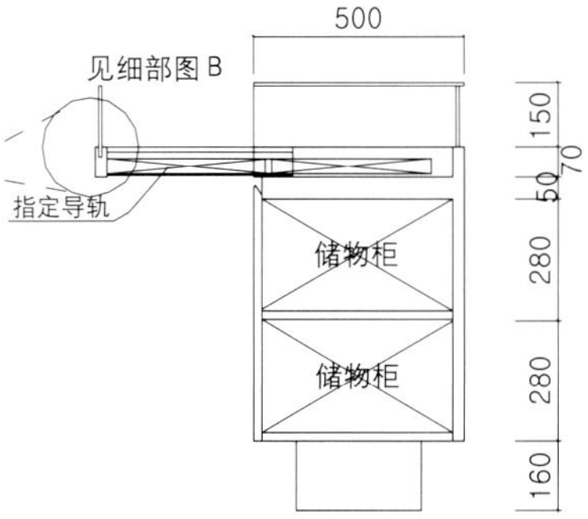

剖面图

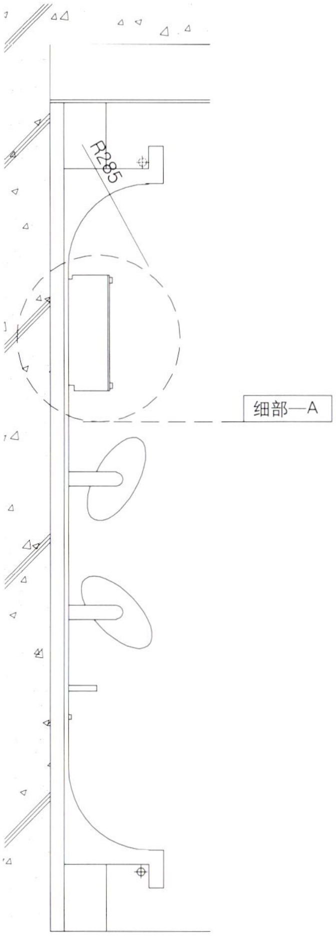

剖面图

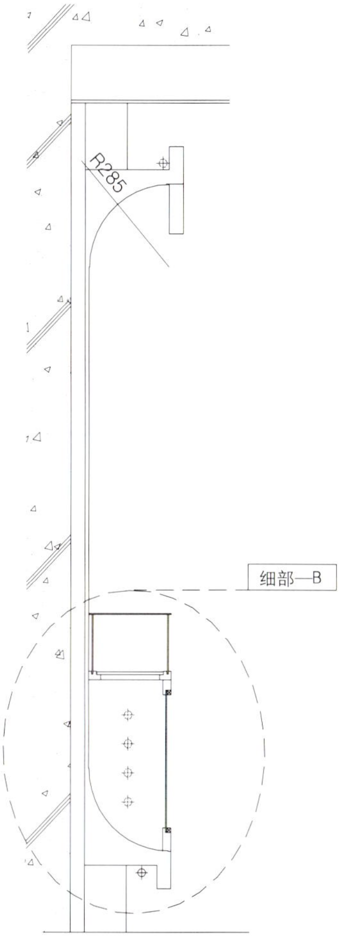

剖面图

■ 一层展示区

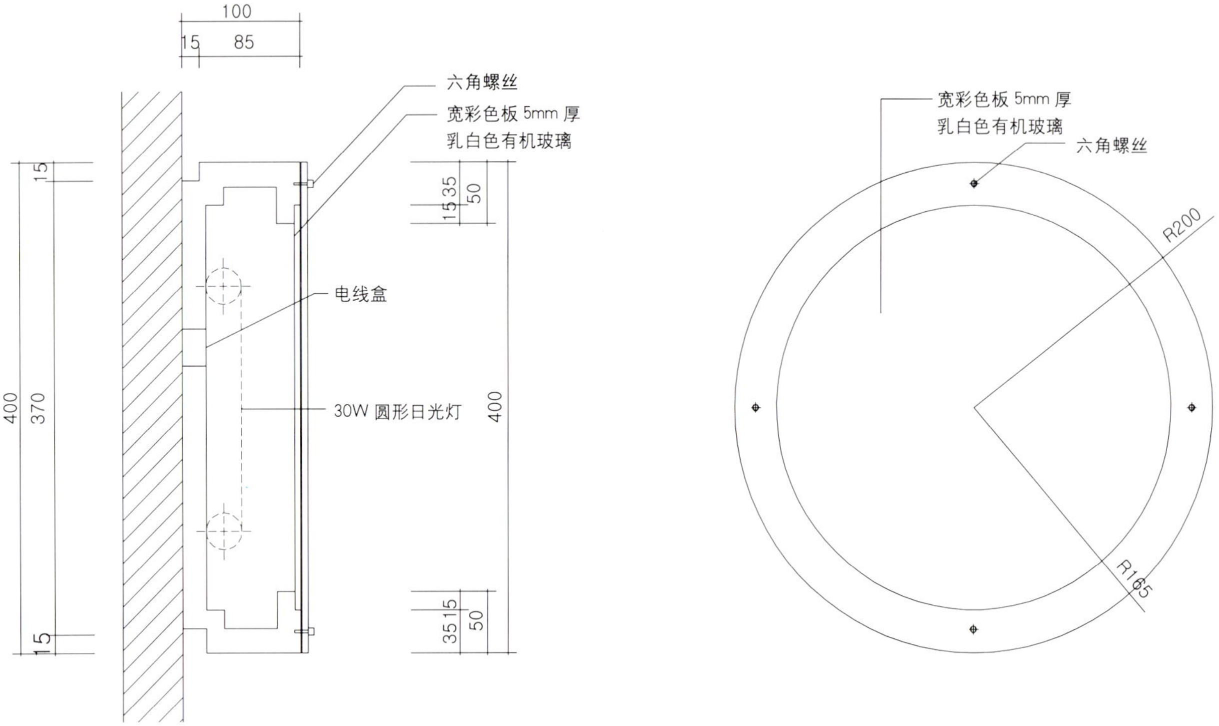

细部图 A—剖面和正视图

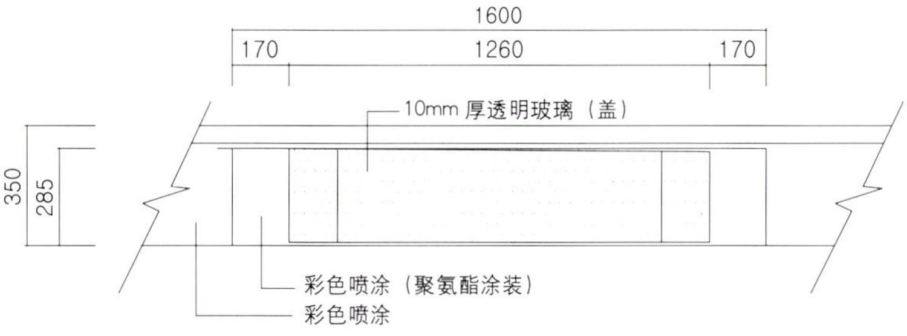

细部图 B—顶视图

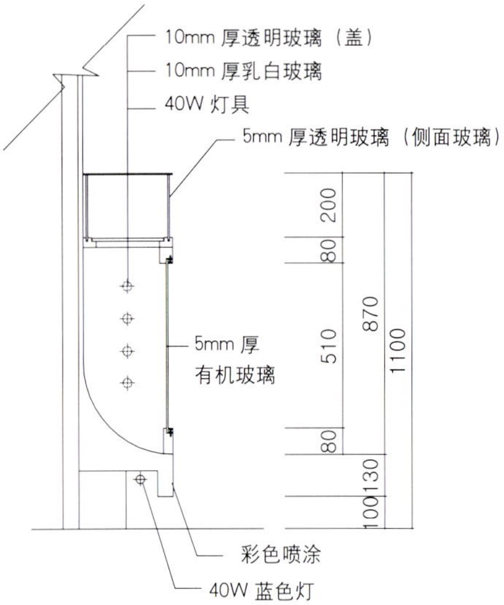

剖面图

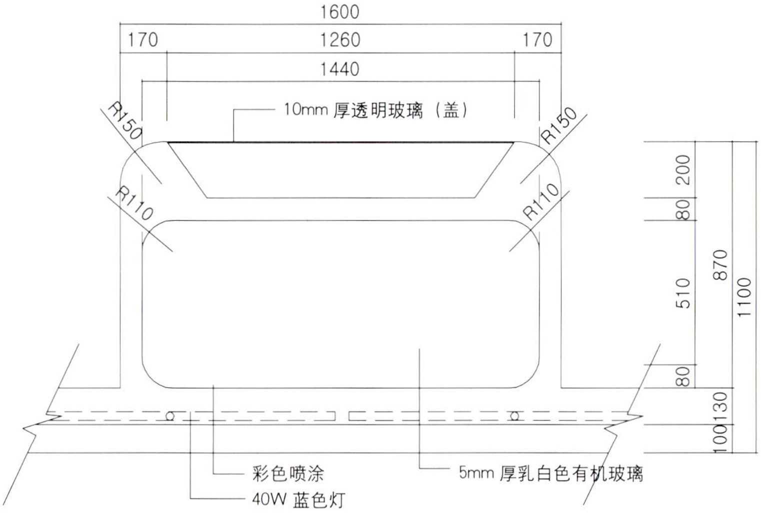

正视图

■ 一层演示区

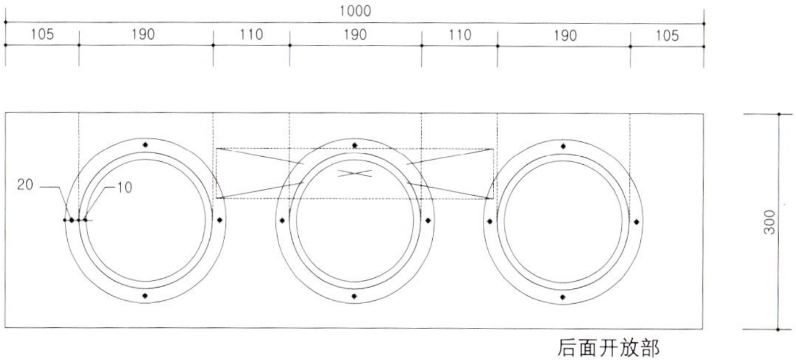

正视图

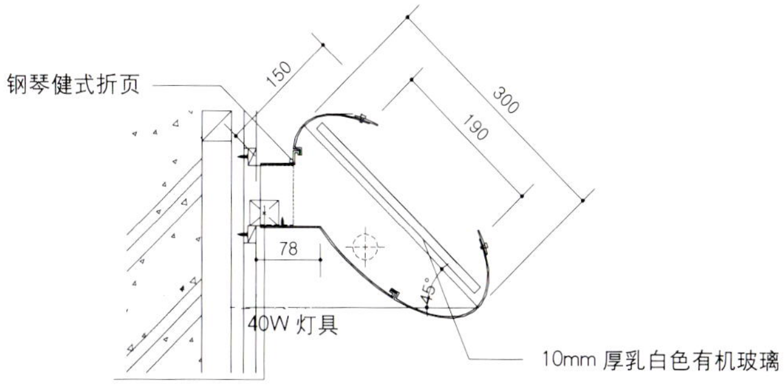

剖面和细部图

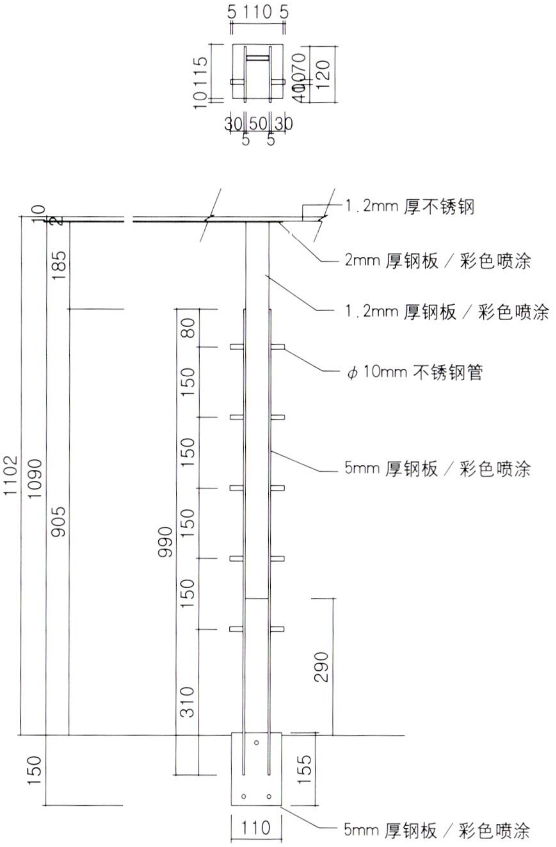

（楼梯）扶手一正视图

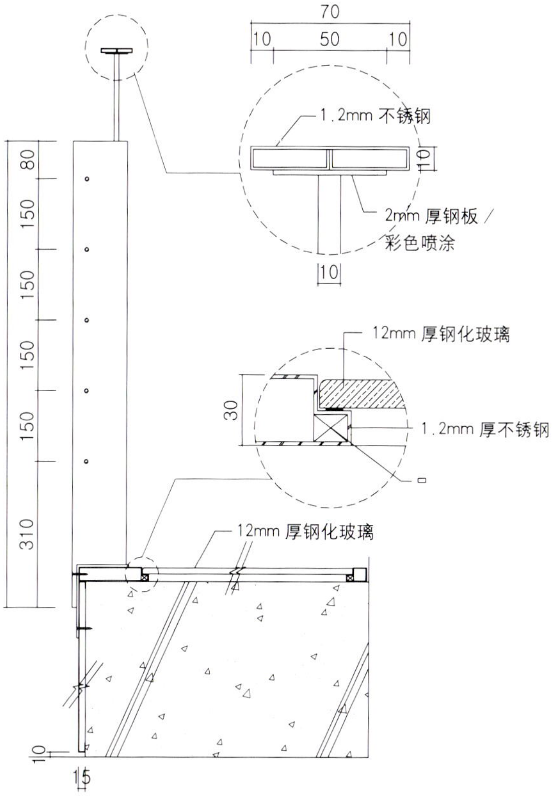

（楼梯）扶手一侧剖面图

■ 灯光箱

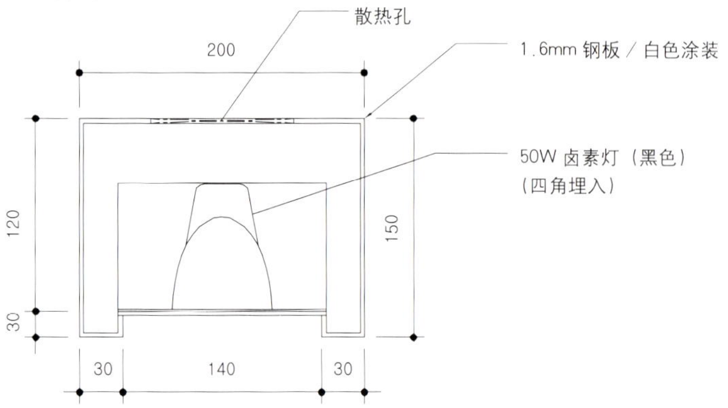

细部图

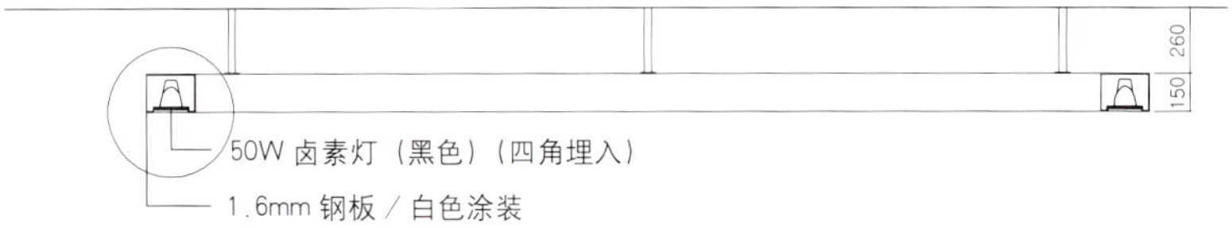

正视图

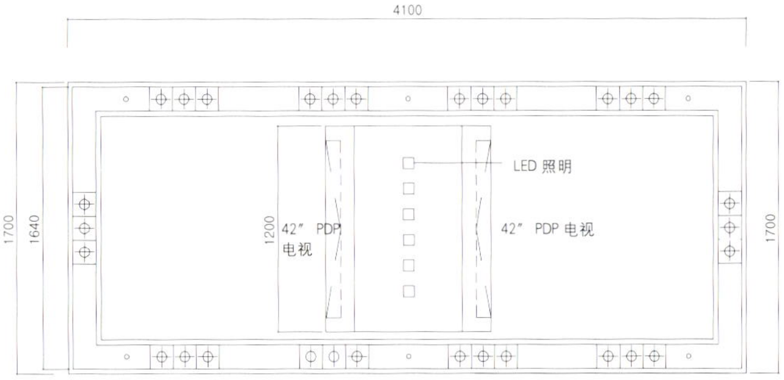

顶视图

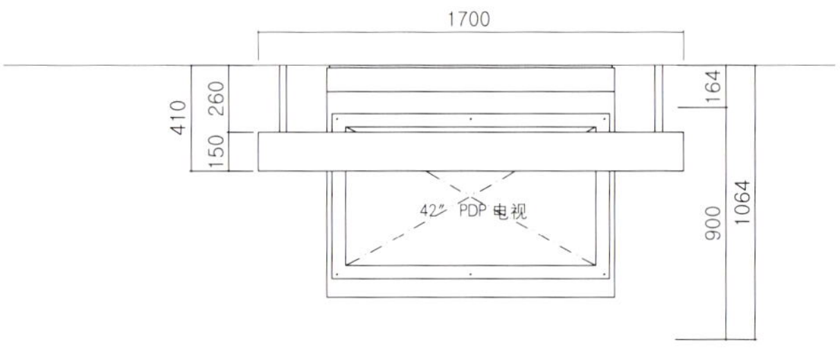

正视图

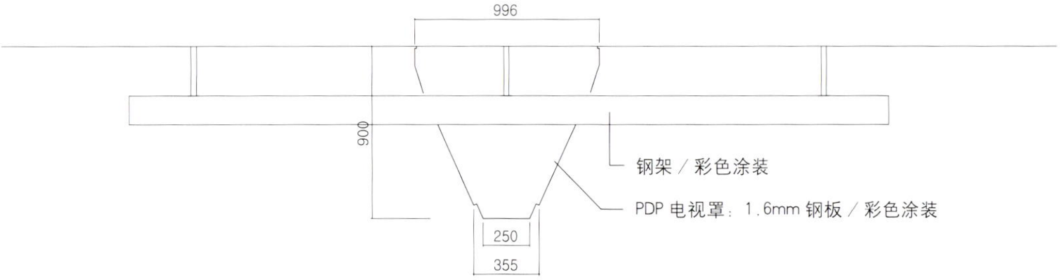

侧视图

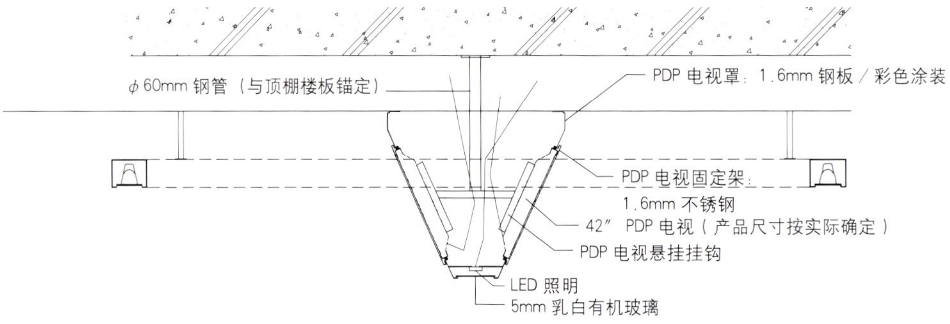

剖面图

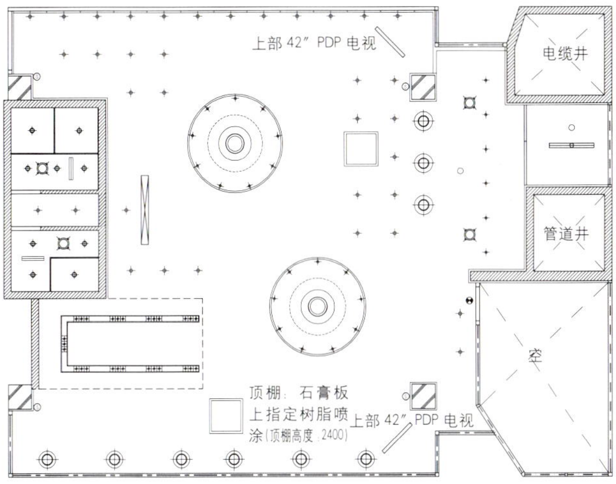

二层顶棚平面图

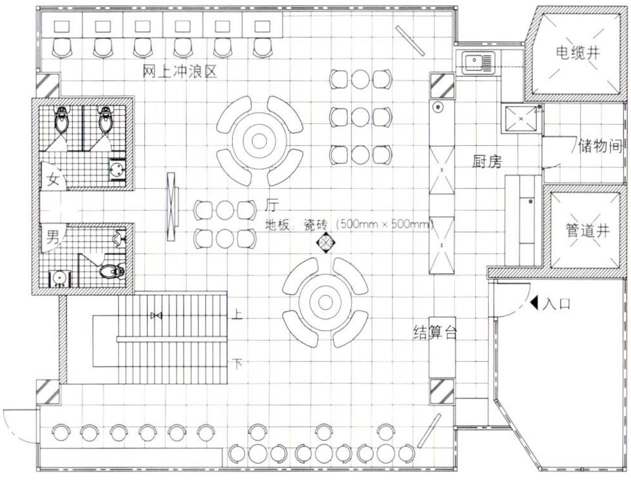

二层平面图

■ 二屋柱子

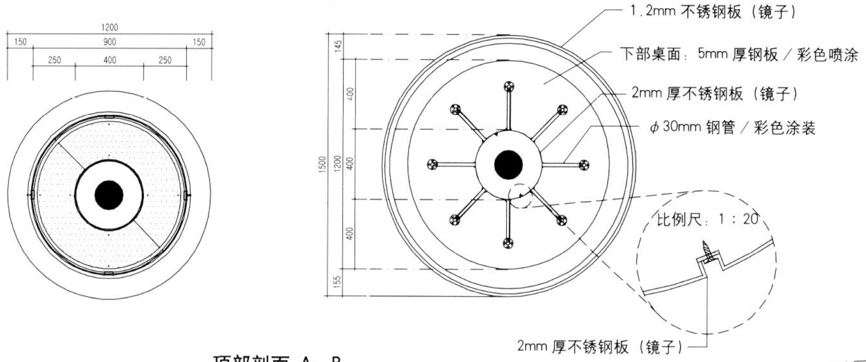

顶部剖面 A、B

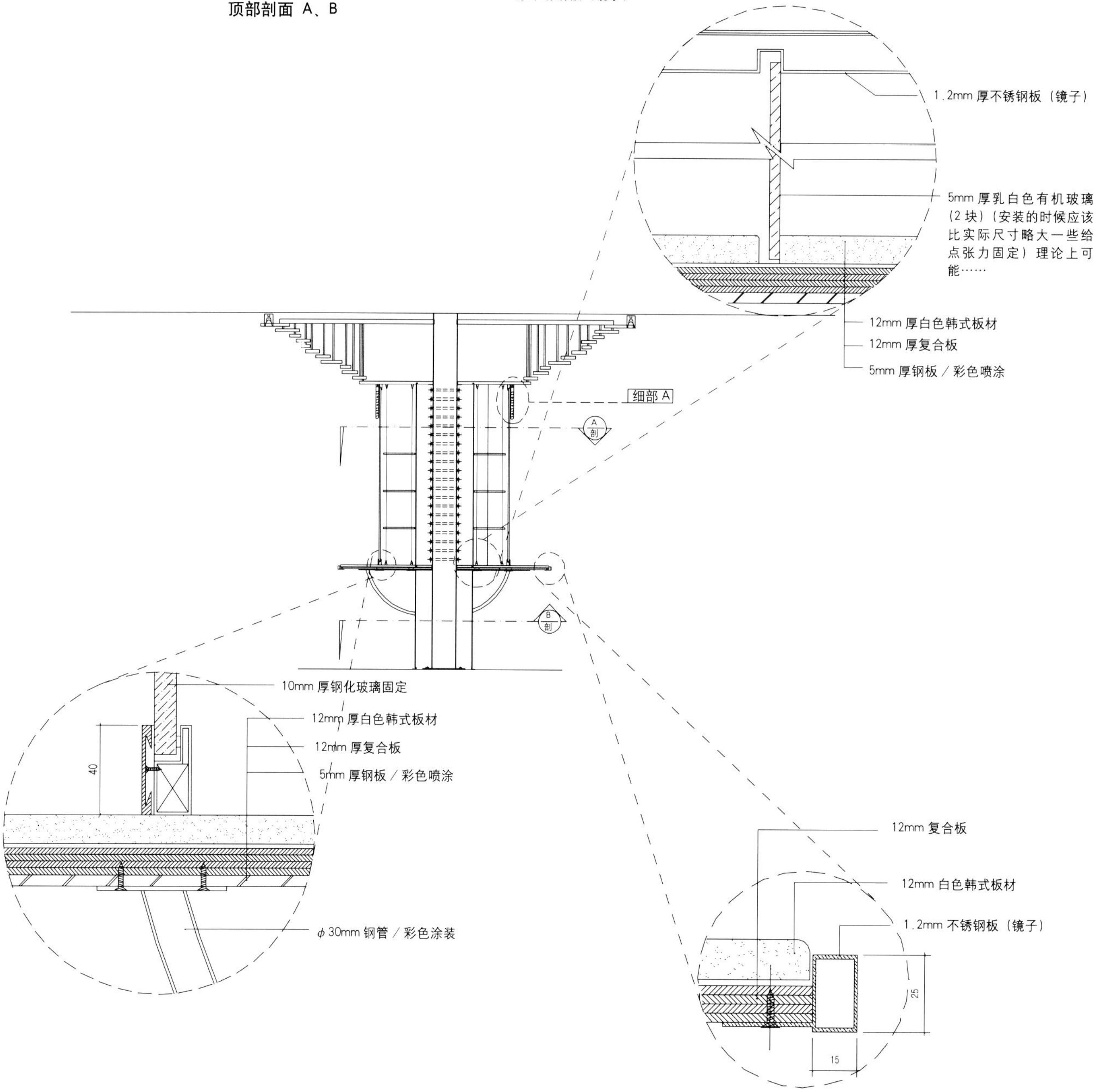

剖面和细部图

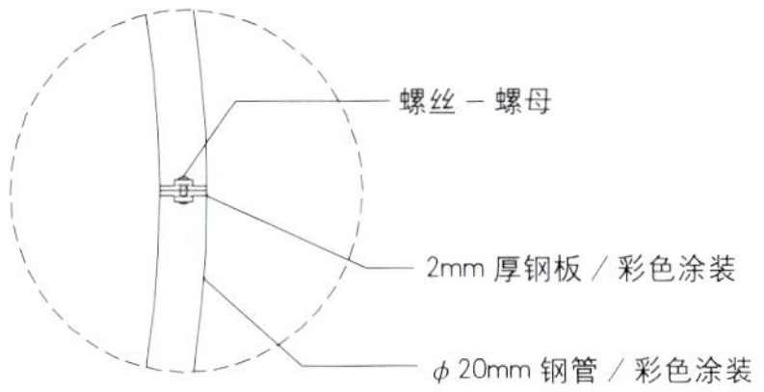

剖面一顶视图

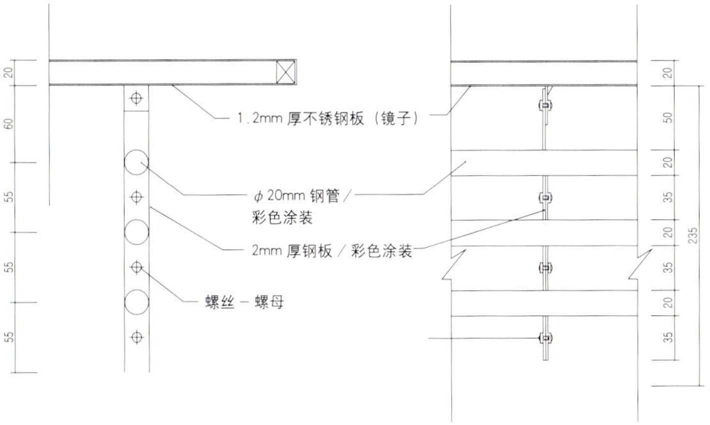

剖面和细部图

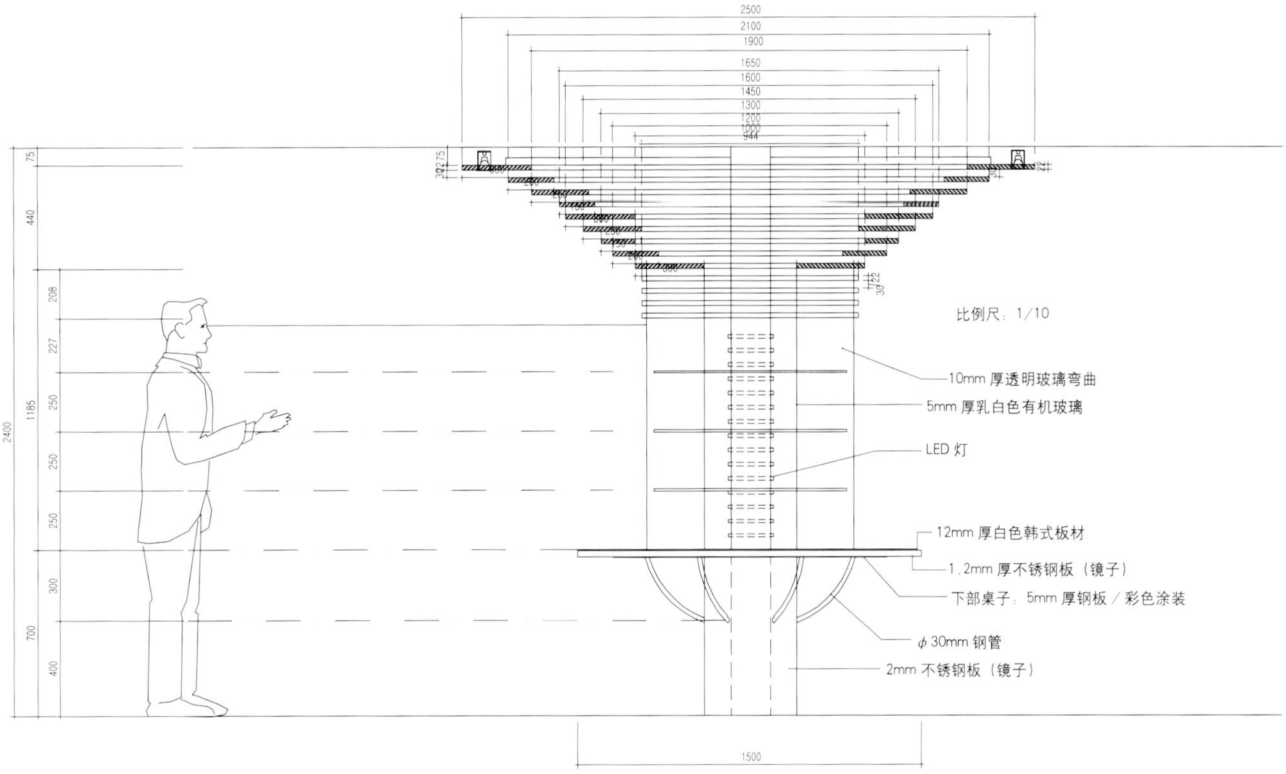

剖面图

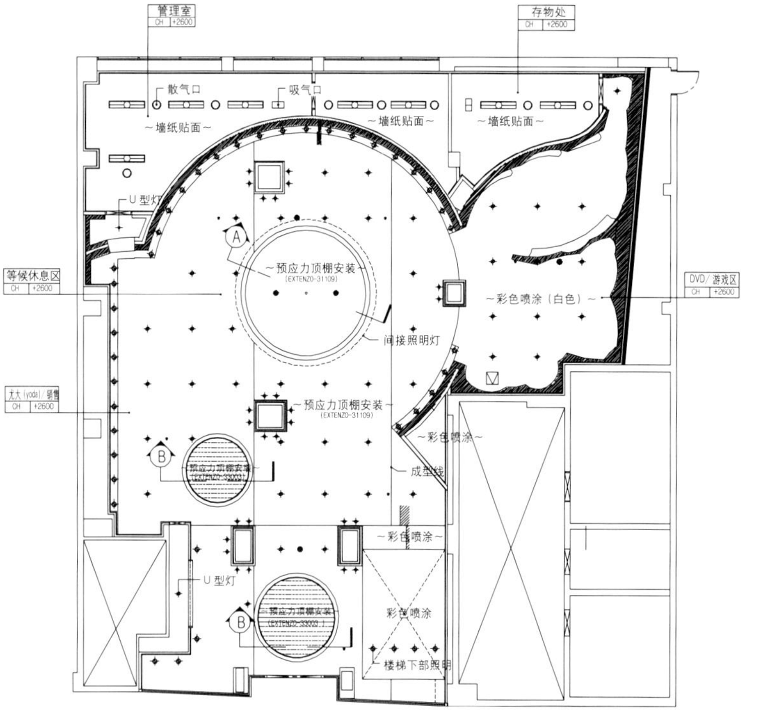

一层顶棚平面图

◪ 图例

符号	名称和规格	数量 (EA)
✦	CDM（厅）	46 EA
✦	AR111（顶棚、隐藏灯光）	42 EA
✦	射灯	26 EA
✦	U 型灯 （13W×2）	1 EA
⊟	埋入 40w 灯具×2	8 EA
——	间接照明灯	
○	散气口	8 EA
◫	吸气口	2 EA
●	扬声器	5 EA
•	热感应器	5 EA
•	烟感应器	1 EA
⊠	检查口	1 EA
•	监视摄像头	1 EA

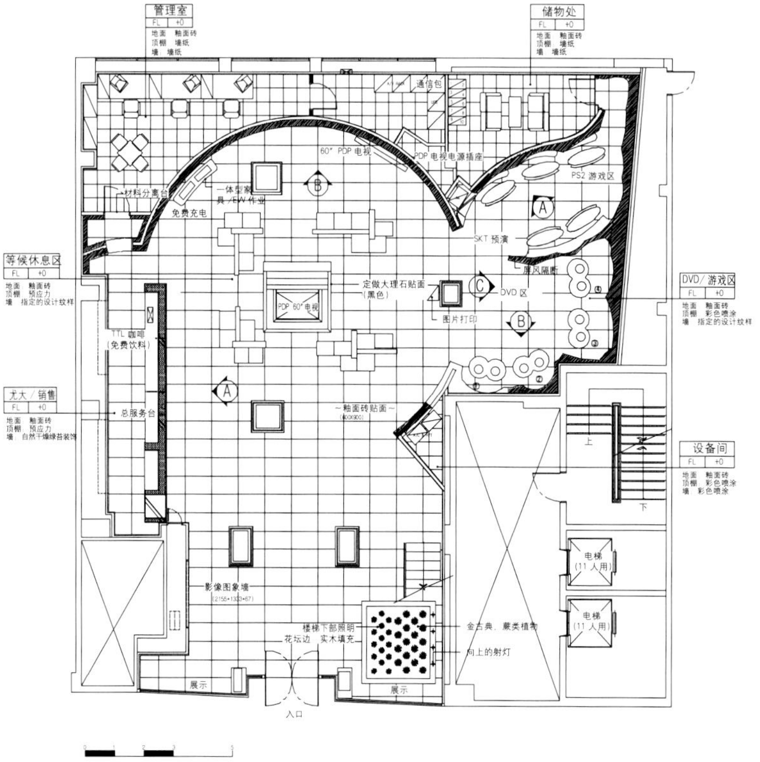

一层平面图

企划：（株）TBWA 韩国设计公司 | **设计：** Multi & Max 设计公司 文锦、金在恩、金胜焕（02）547-7507 | **施工：** Multi & Max 设计公司 金荣焕、金龙哲、郑庆允 | **地址：** 釜山市福山镇区釜前洞 241-38 | **面积：** 697m² | **装修：** 地面／釉面砖、实木 墙／原木、EPS 模块、风干苔藓 顶棚／预应力顶棚 | **摄影：** 金明植

Planning: TBWA ₩ KOREA Co., Ltd. | **Design:** Multi & Max · Moon, Jin / Kim, Jea eun / Kim, Seung hwan | **Construction:** Multi & Max · Kim, Young hwan / Jin, Yong cheol / Jeong, Keong yun | **Built Area:** 697m² | **Finish :** Floor / Polishing Tile, Solid Wood Wall / Lumber, EPS Modeling, Dry Moss Ceiling / Extenzo (Stretching Ceiling) | **Photograph:** Kim, Myung sik

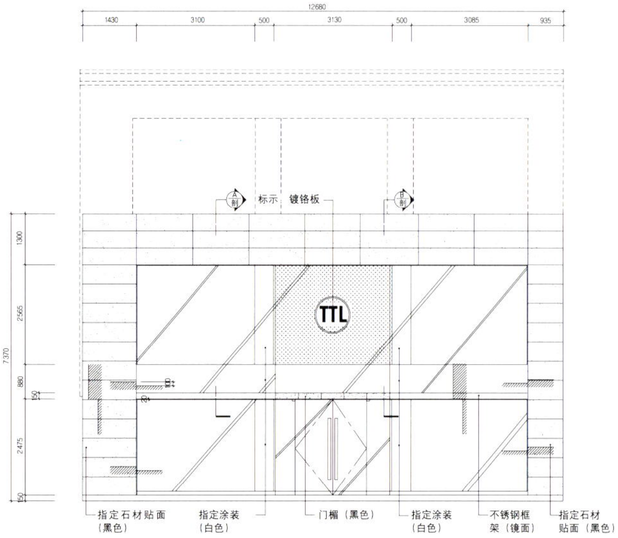

正视图

■ 正立面图

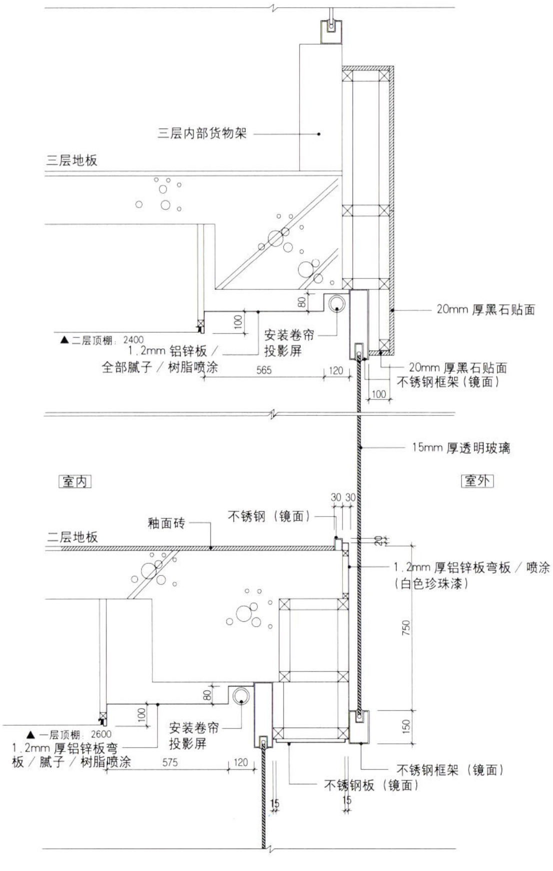

剖面图 A

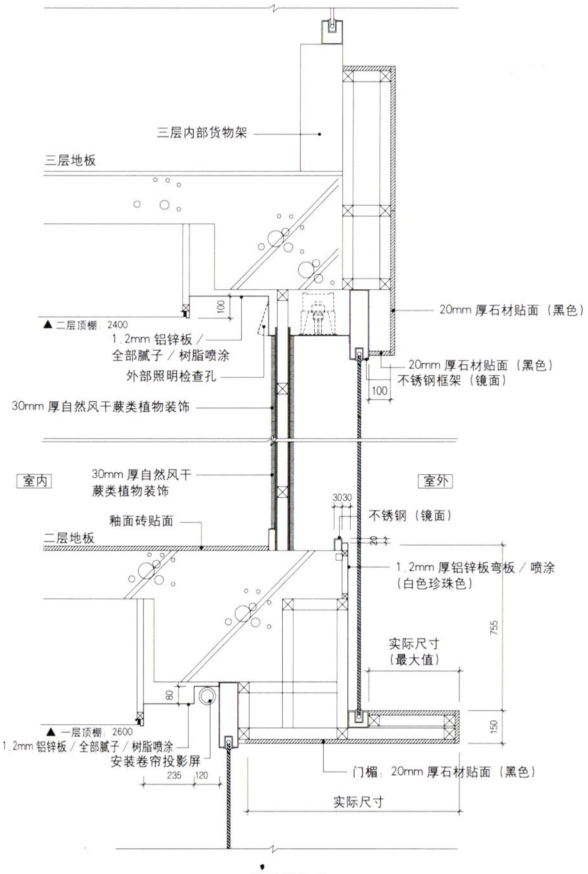

剖面图 B

最近的流行趋势是采用“健康生活价值”的材料和设计，比较有代表性的是 TTL 歌区，它是以白色和绿色为主色调，消费人群锁定在 20 岁左右的年轻一族。

“Natural,Nobless,Nomad”的 3N 设计主题和所谓的健康社会流行趋向连在一起，通过人的五官感受到舒适的空间效果。为了增加手持 TTL 设备的新鲜感，提升新潮领导者的形象，室内装饰材料使用了天然的风干苔藓，台阶采用了弯曲的原木，还使用了 EPS 模型等新材料。为了解决过矮的顶棚问题使用了预应力顶棚。

另外，位于一层休息区的数字莲池把水和数字结合在一起，试图采用新方法把自然要素（水）和尖端技术（数字）融合起来。今后 TTL 歌区的目的就是要为游玩时间极度缺乏的青少年及年轻人提供舒适的休息地。

■ 一层等候休息区

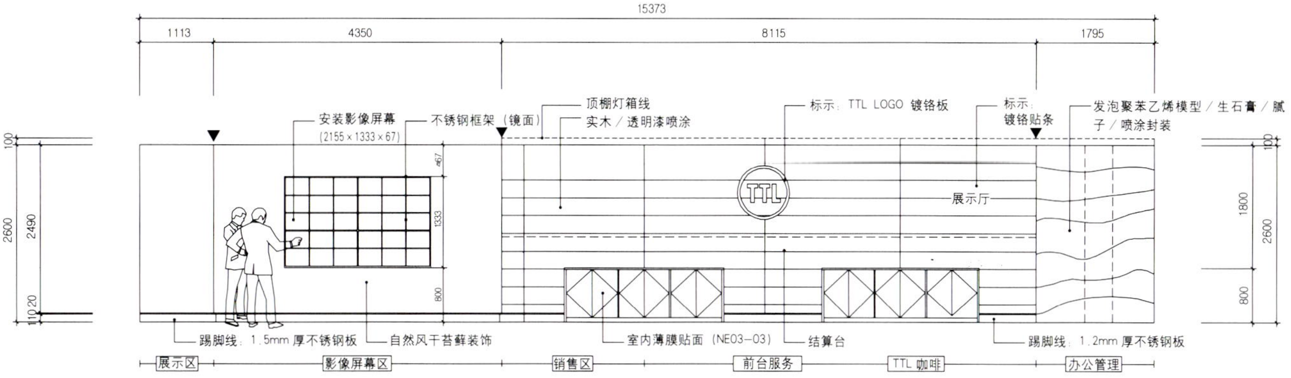

立面图 A（单位：mm）

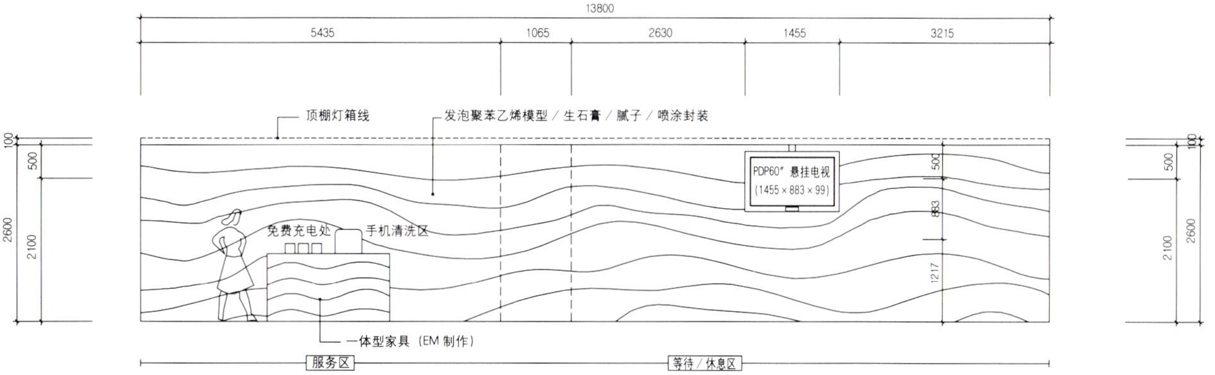

立面图 B（单位：mm）

■ 一层数字媒体电视

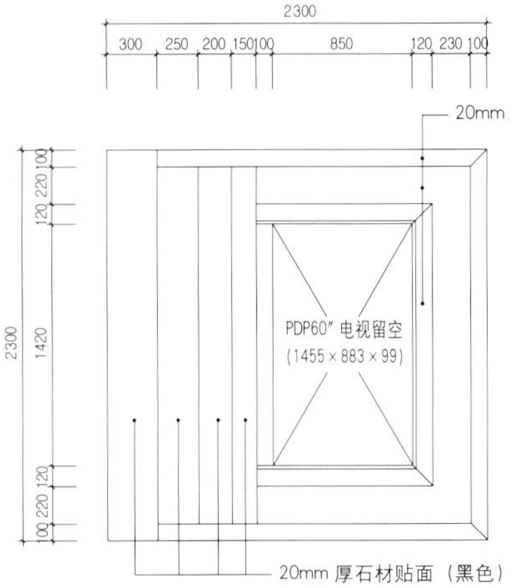

顶视图（单位：mm）

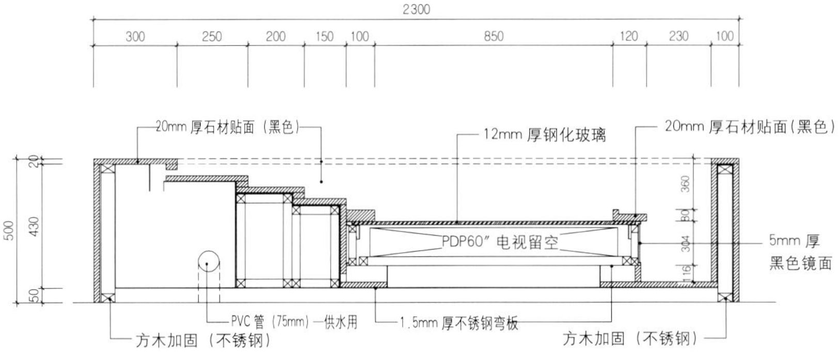

剖面图

■ 一层尤大结算台

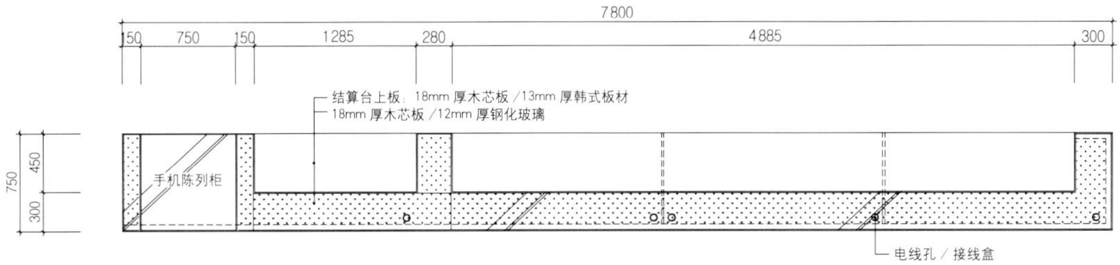

顶视图

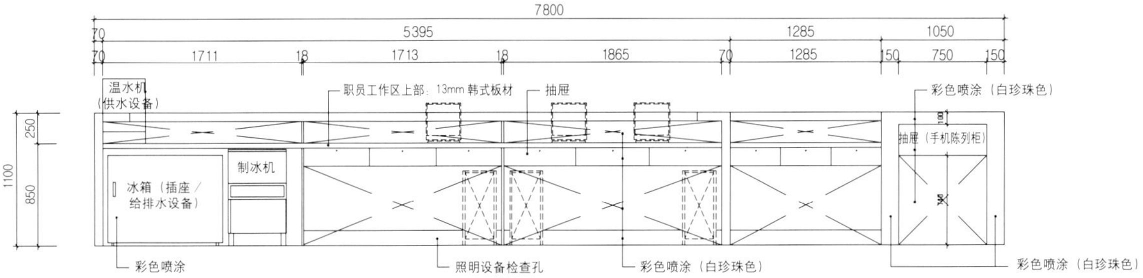

后视图

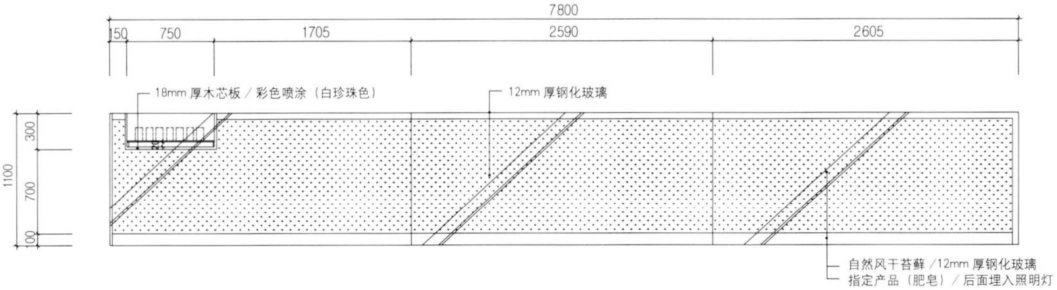

正视图

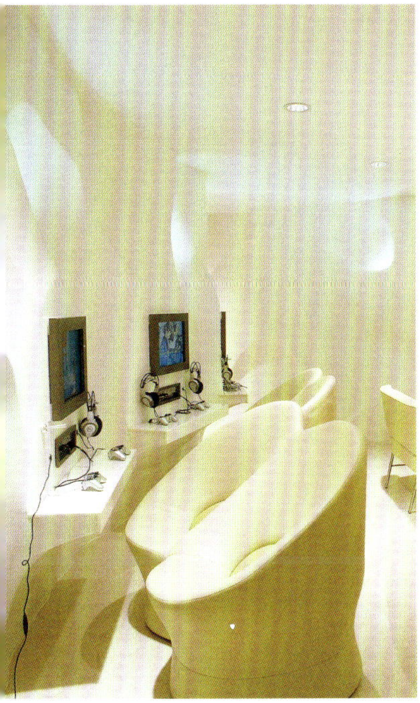

■ 一层DVD/游戏区

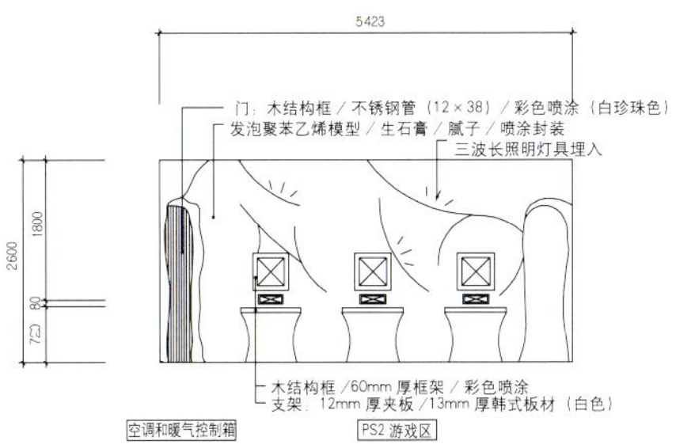

立面图 A

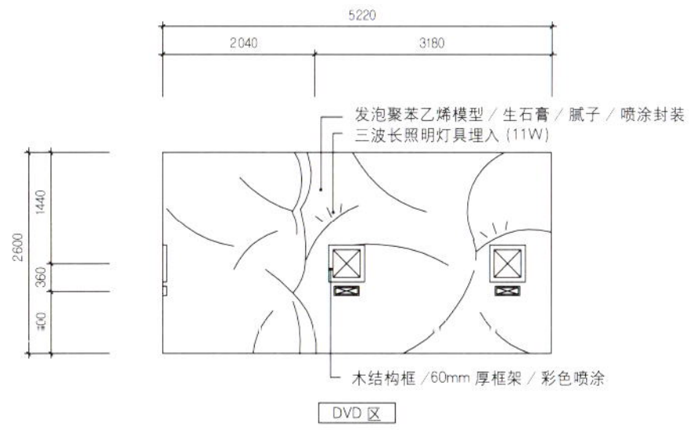

立面图 B

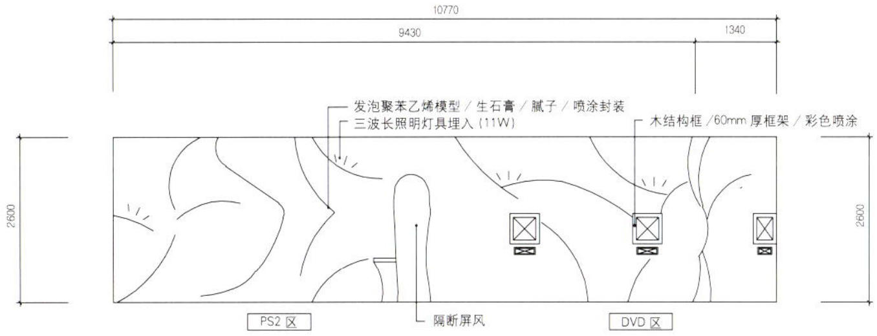

立面图 C

■ 一层墙面细部

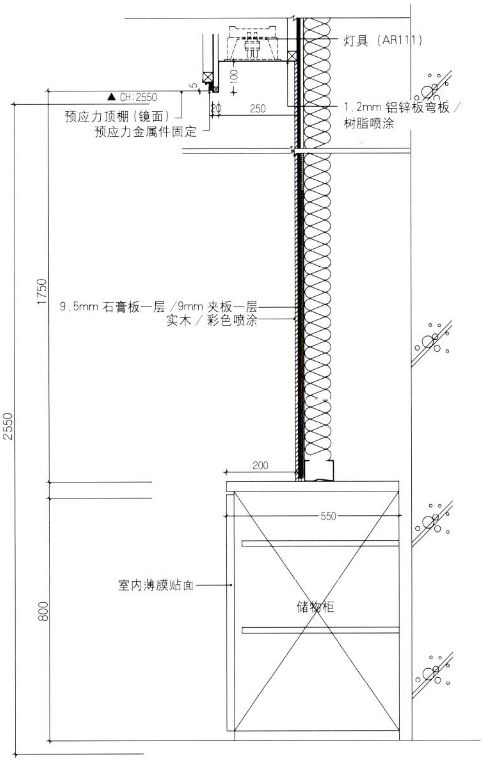

尤大区—剖面图

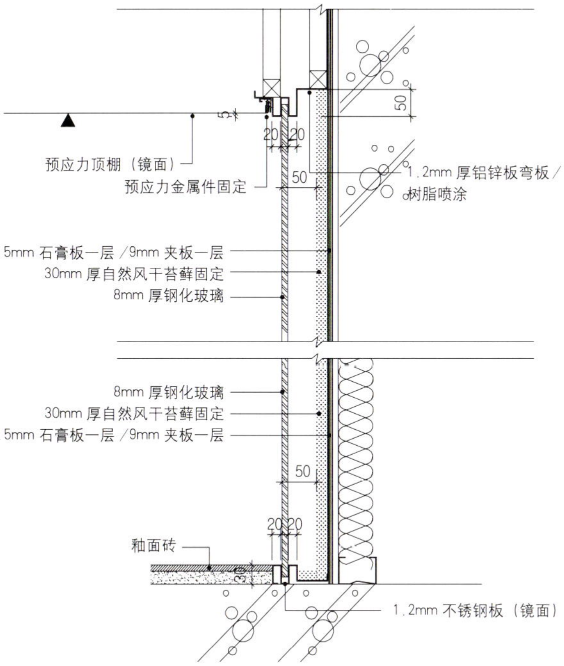

Moss 侧玻璃—剖面图

The fresh atmosphere of TTL ZONE effectively exhibits its brand image that targets 20s with material and design reflecting the recent trend 'Well-being', and finish materials in bright, cheerful white and green.

Design concept Natural Nobless Nomad and social trend 'Well-being' were engaged to present a comfortable space where one can feel nature through the five senses. New materials, such as dry moss, curved lumber staircase and EPS modeling, were used to emphasize freshness and the image of a trend leader. Extenzo Stretch ceiling solves the problem of low ceiling height.

Also, a digital pond of the 1st floor waiting area introduced both water and a digital material, making an attempt of harmonization of nature and technology. I expect TTL ZONE to become a relaxing space to teenagers and young people, of whom the entertainment culture is in severe shortage.

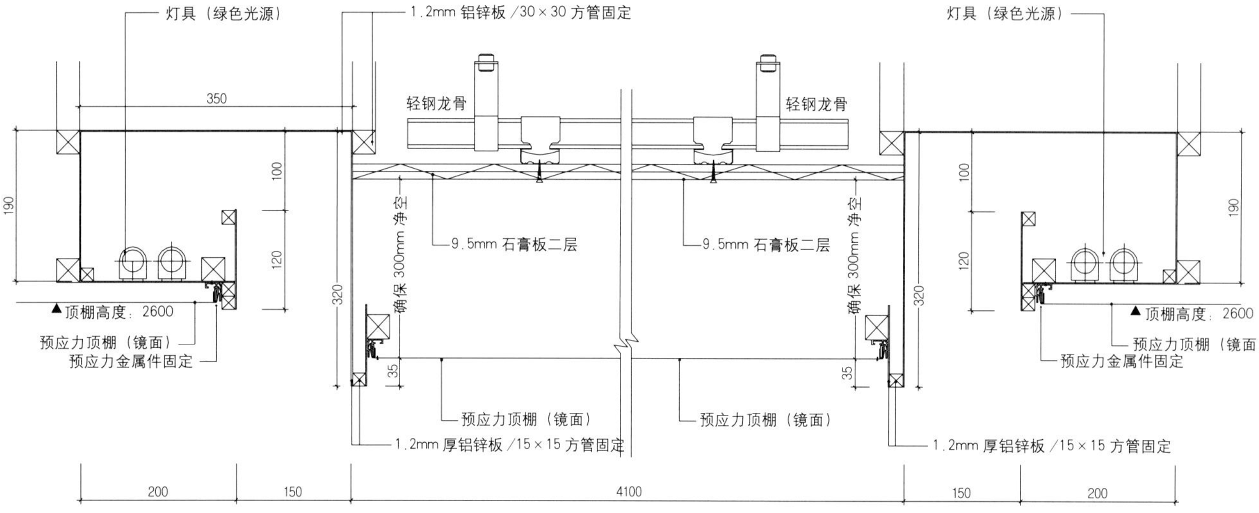

一层主灯箱一剖面图（单位：mm）

■ 楼梯

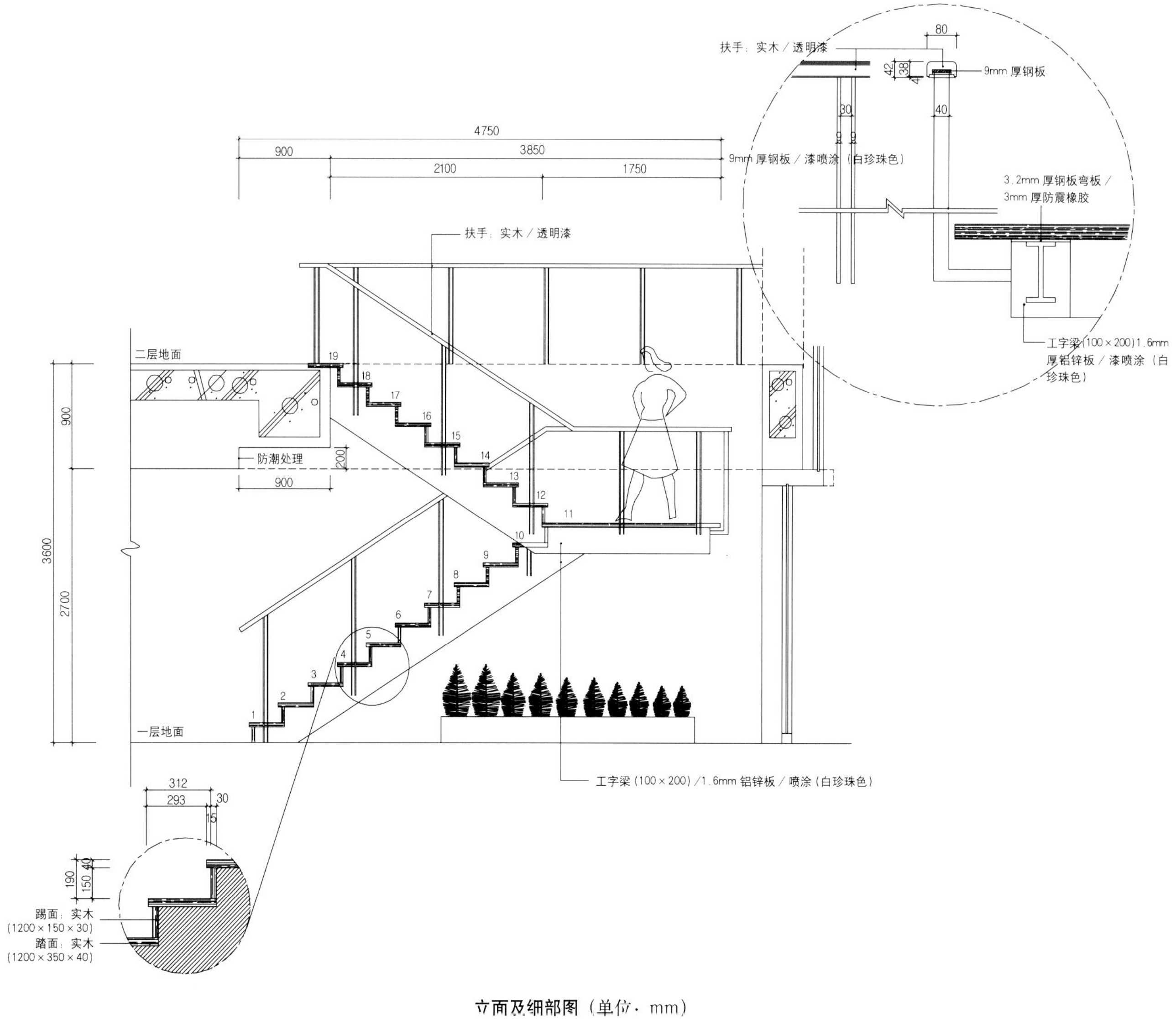

立面及细部图（单位：mm）

TTL
TTL
INFORMATION

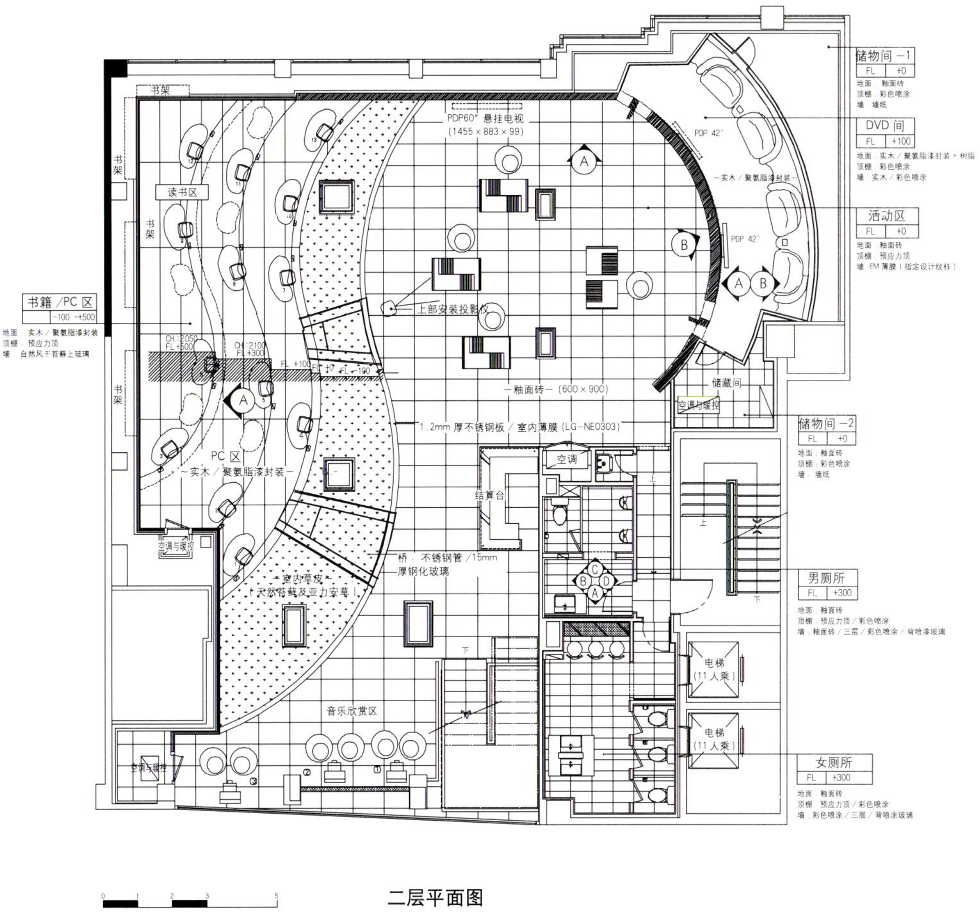
书架
读书区
书籍 /PC 区
PC 区
~实木／聚氨脂漆封装~
空调与缓控
室内草坪
PDP60" 悬挂电视
(1455 × 883 × 99)
上部安装投影仪
~釉面砖~ (600 × 900)
1.2mm 厚不锈钢板／室内薄膜 (LG-NE0303)
桥：不锈钢管 /15mm
厚钢化玻璃
结算台
空调
音乐欣赏区
储物间 -1
DVD 间
酒动区
储藏间
储物间 -2
男厕所
女厕所
电梯
(11 人乘)
电梯
(11 人乘)

二层平面图

■ 二层结算台

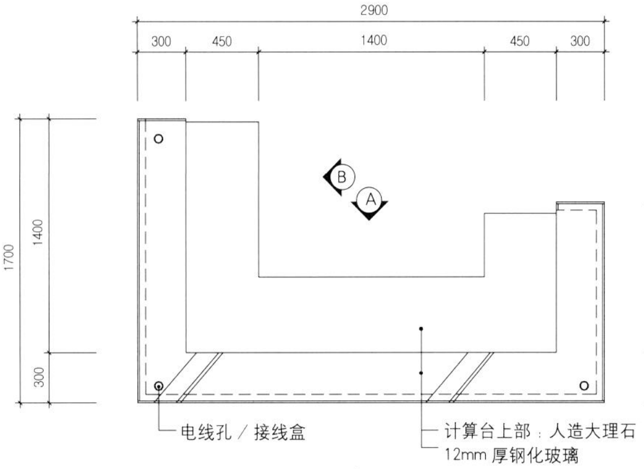

顶视图

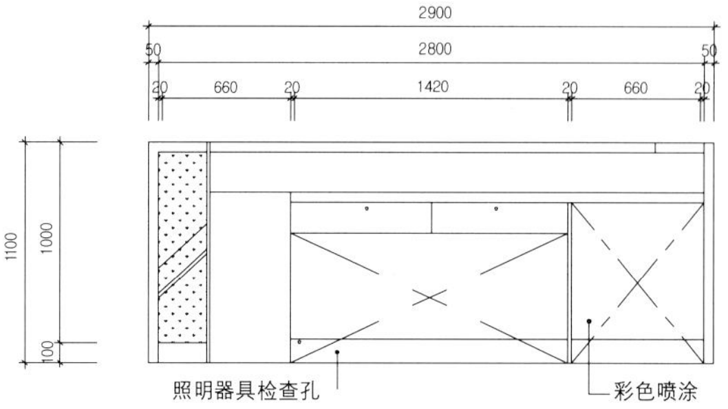

A 向视图

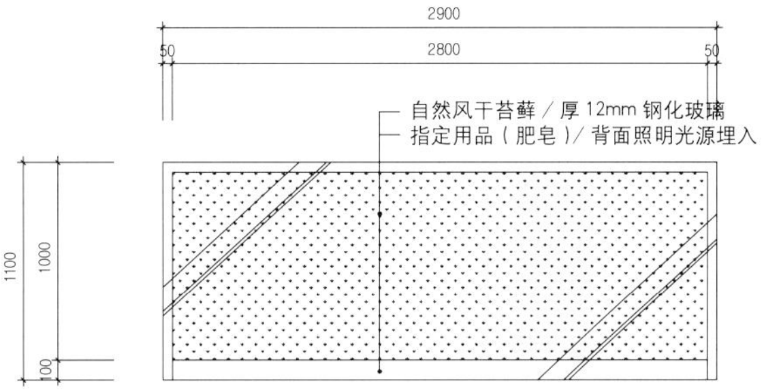

正视图

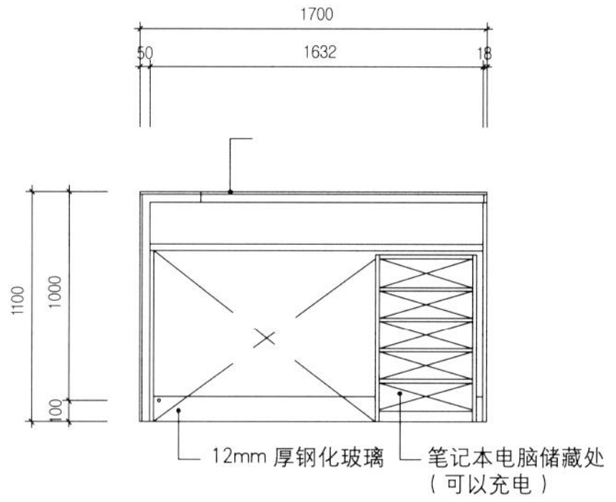

B 向视图

■ 二层活动区

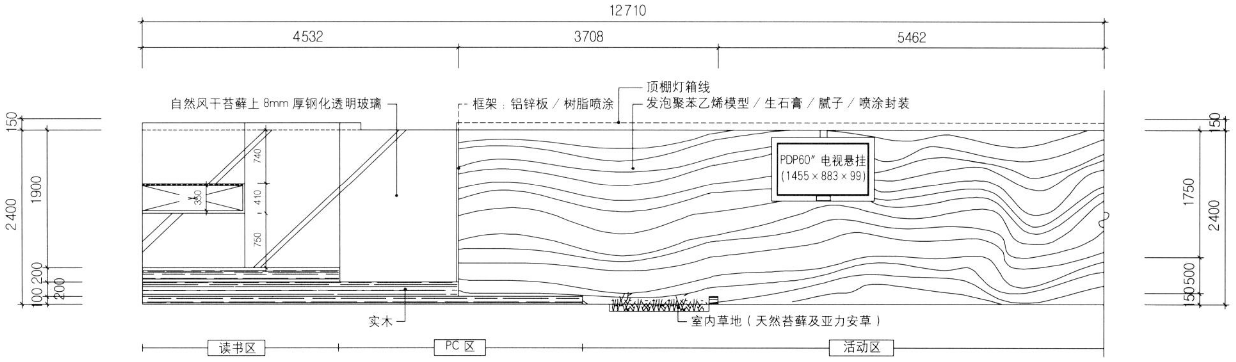

立面图 A（单位：mm）

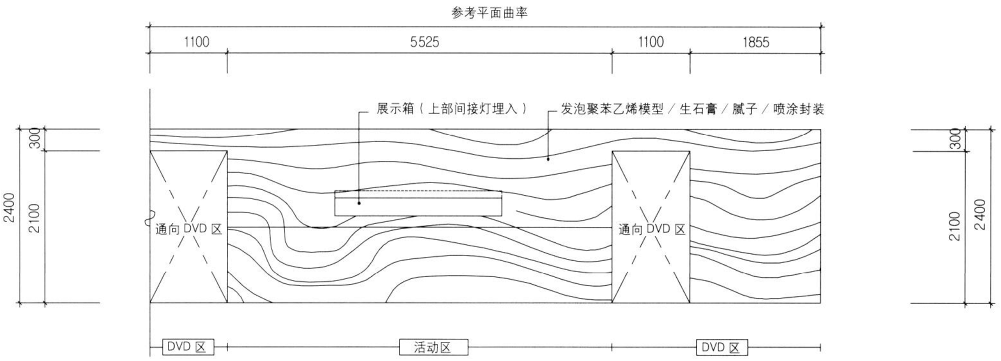

立面图 B

■ 二层活动区—家居

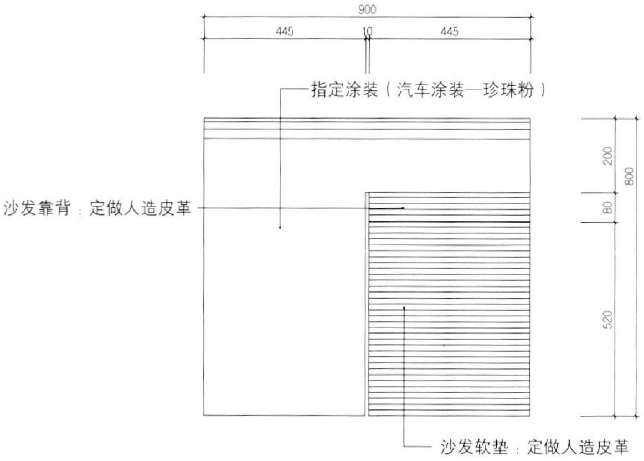

顶视图

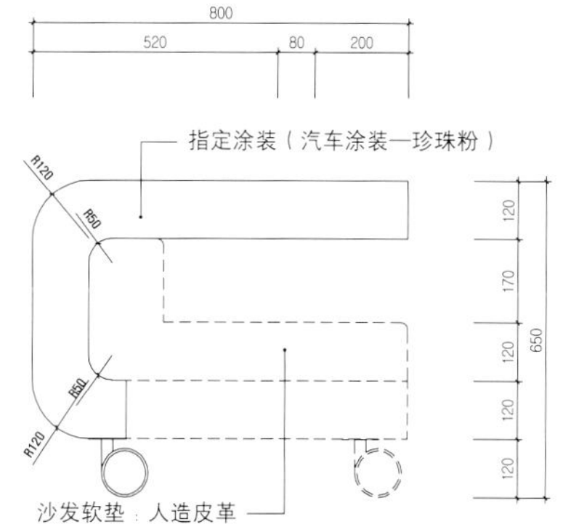

侧视图

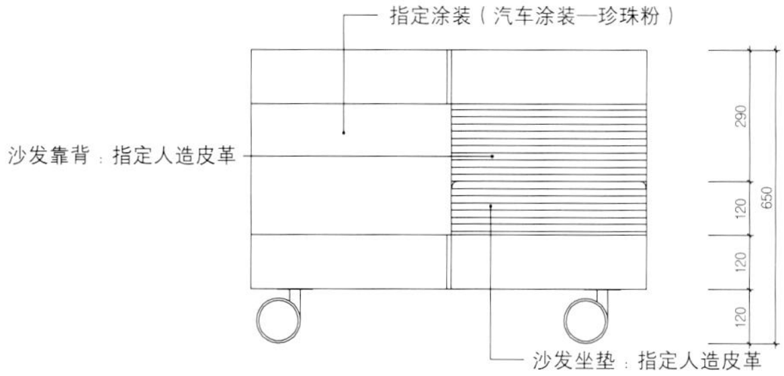

正视图

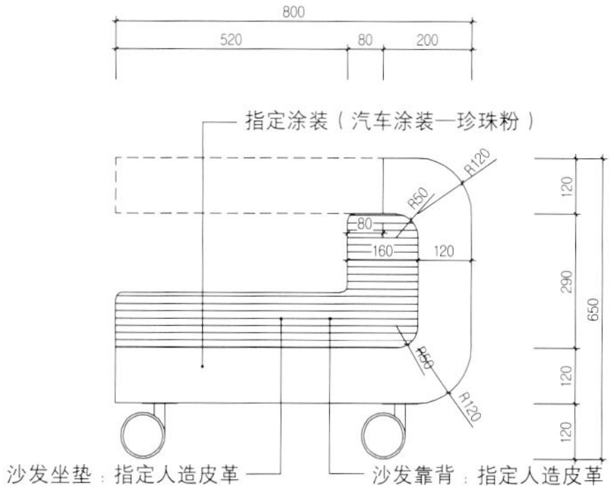

侧视图

■ 二层 DVD区

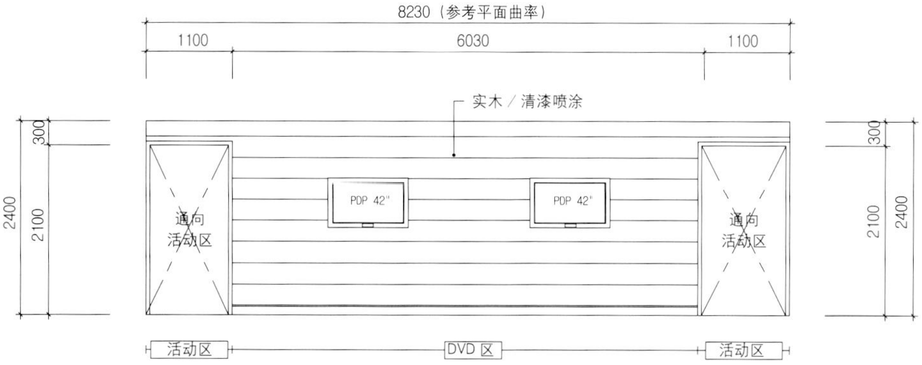

立面图 A

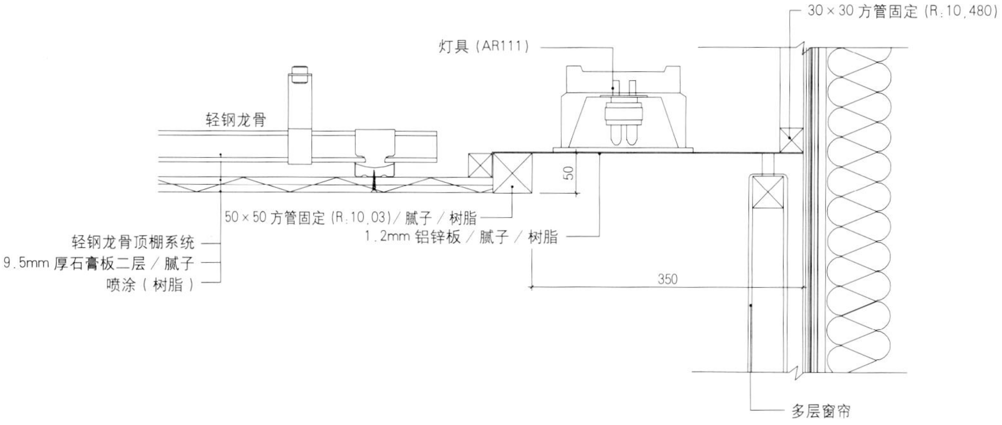

DVD 区灯箱—剖面图（单位：mm）

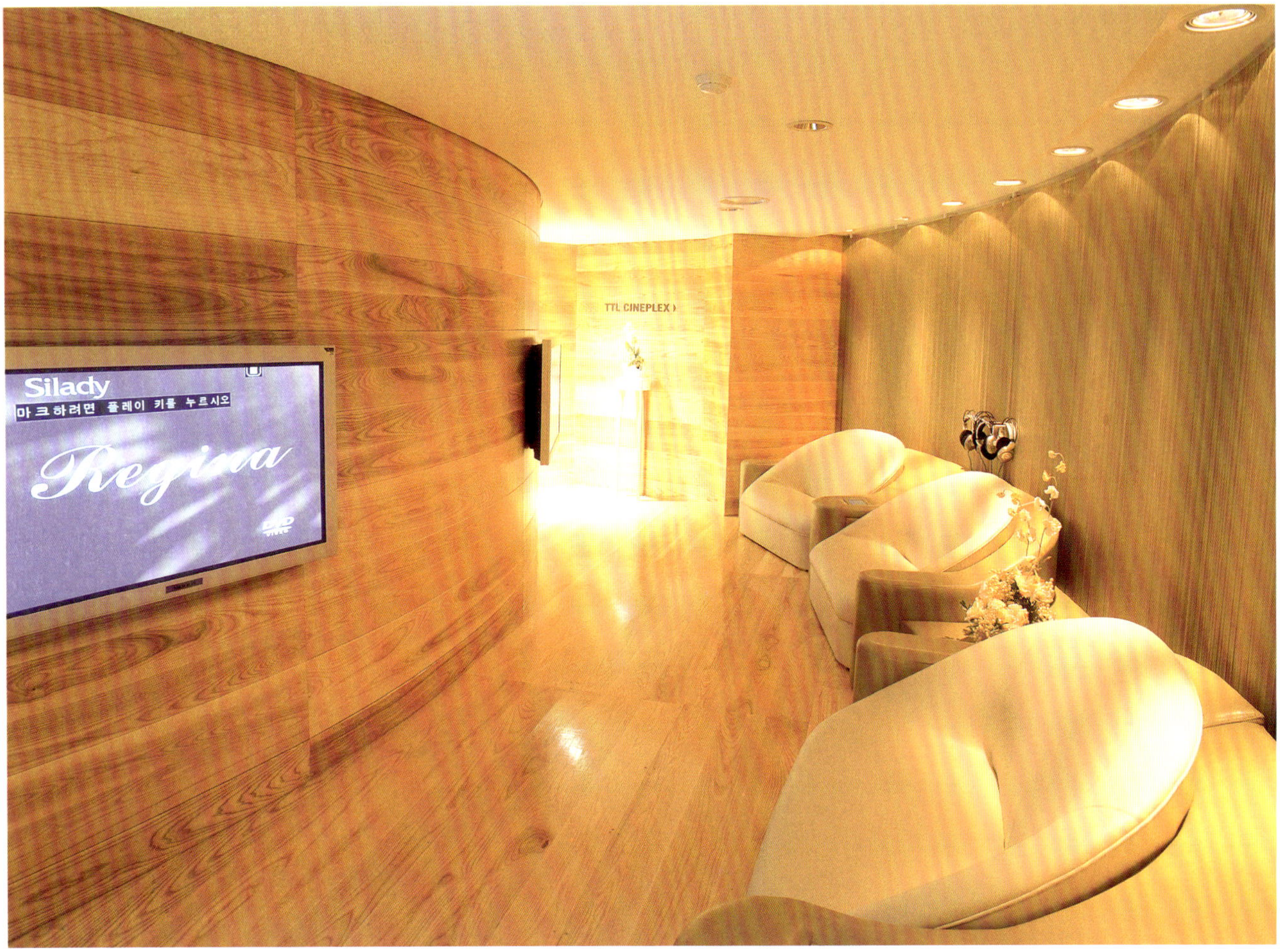

■ 二层DVD区一家具

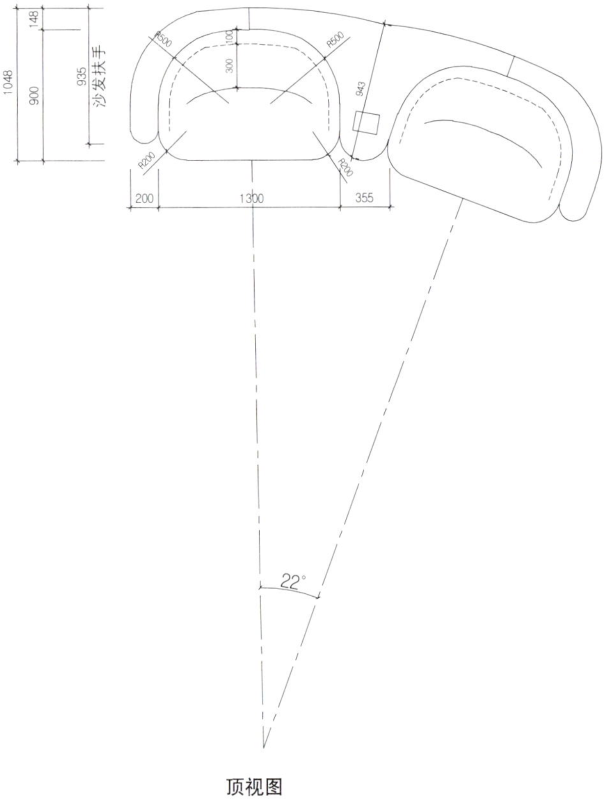

顶视图

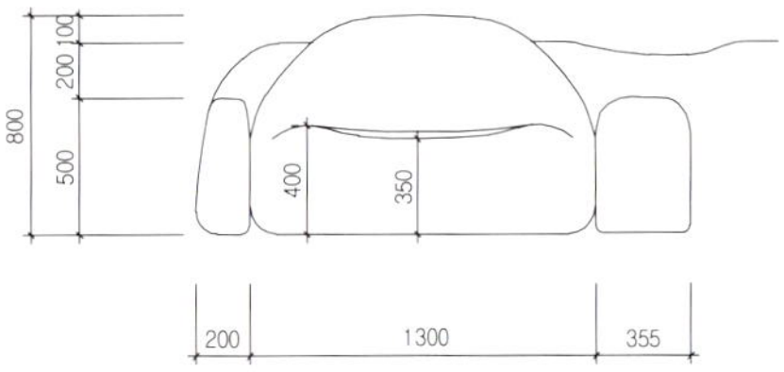

正视图

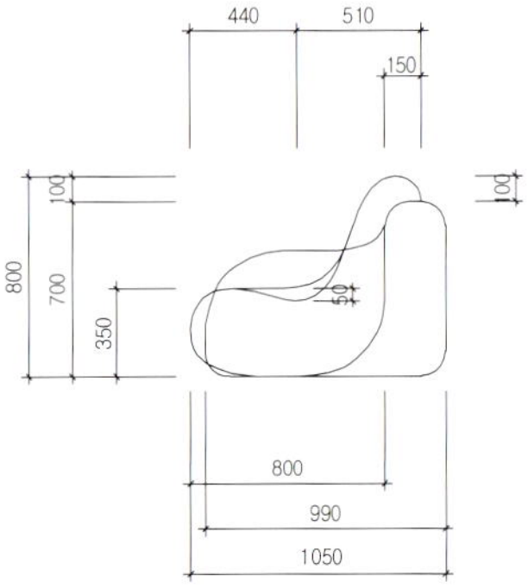

侧视图

■ 二层读书/PC区

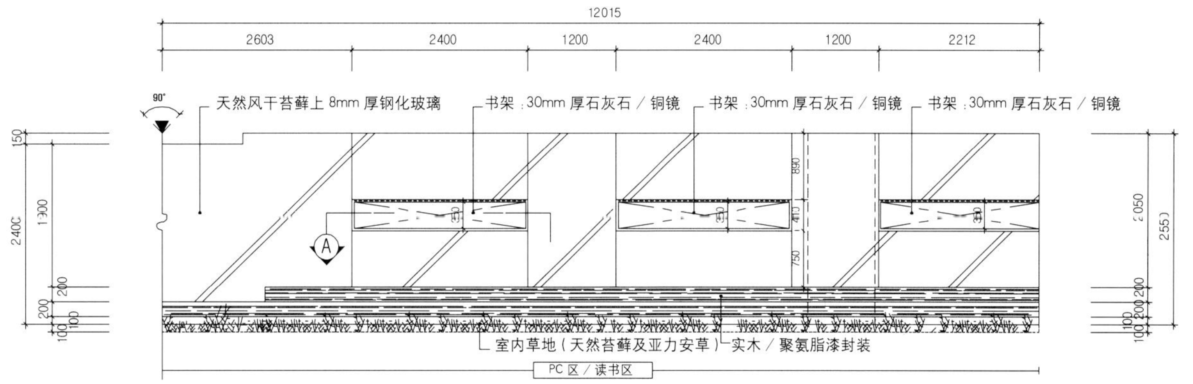

立面图 A

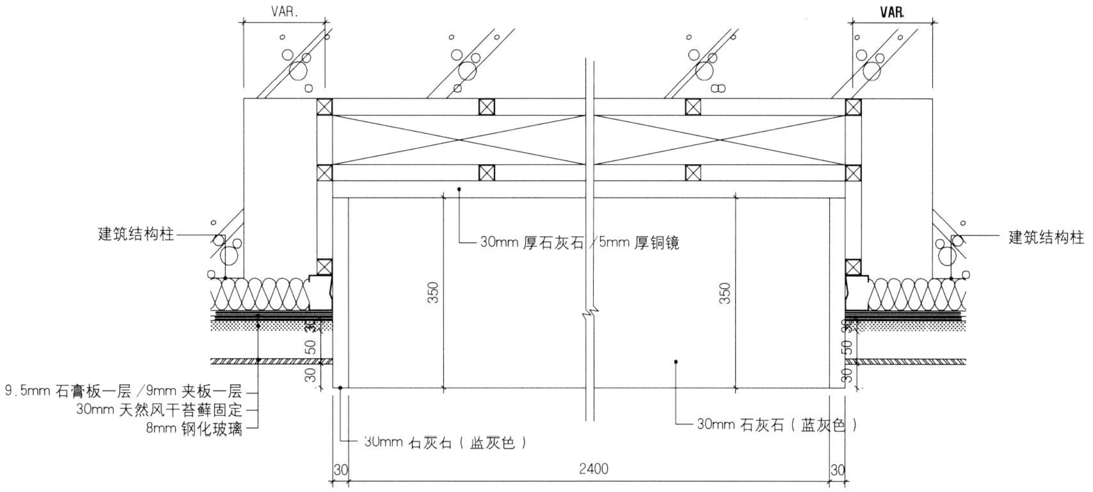

书架—剖面图

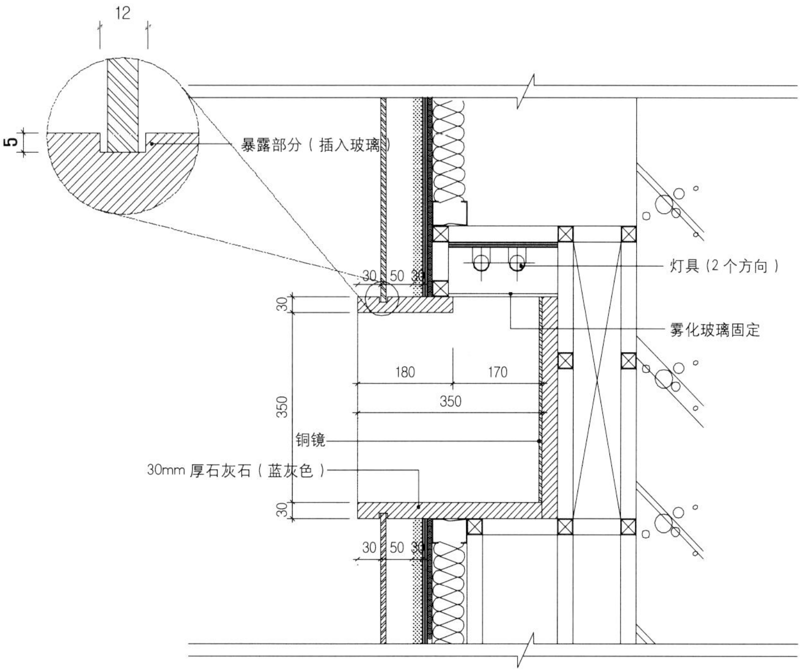

书架—剖面和细部图

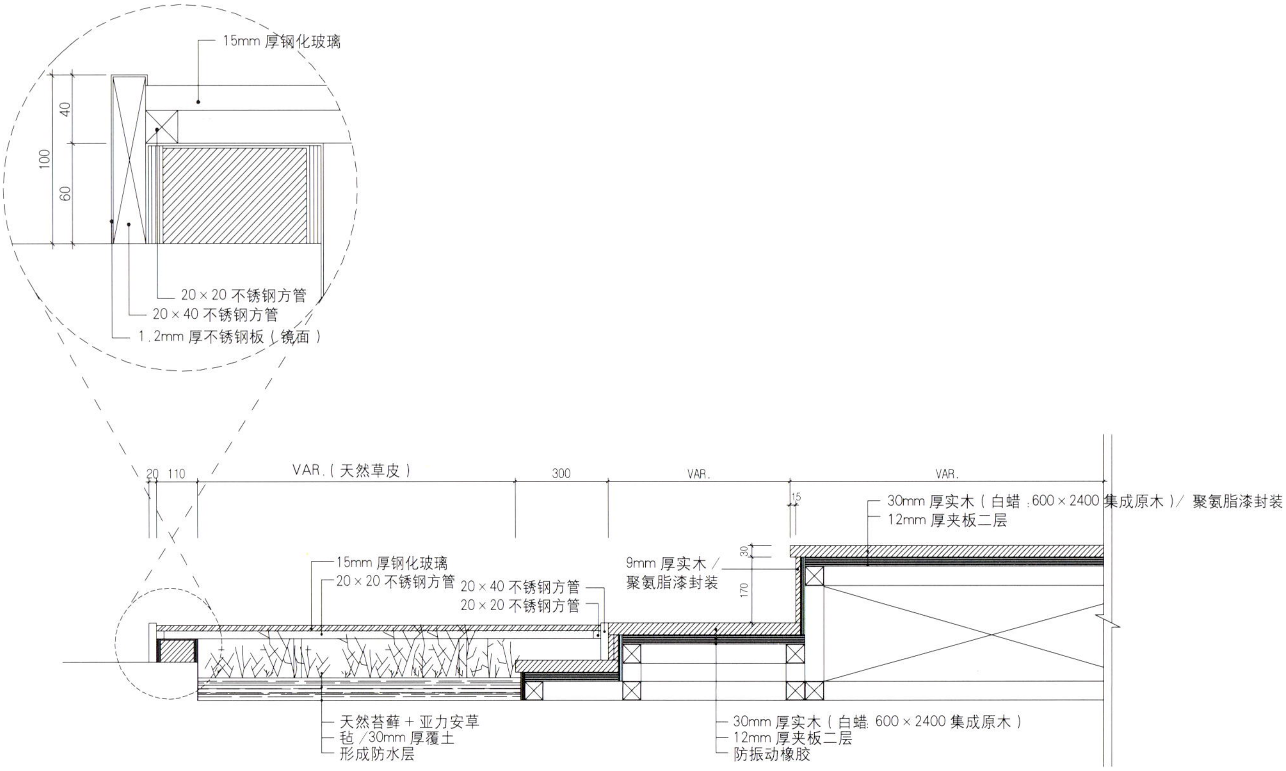

PC 区台阶一剖面图和细部图（单位：mm）

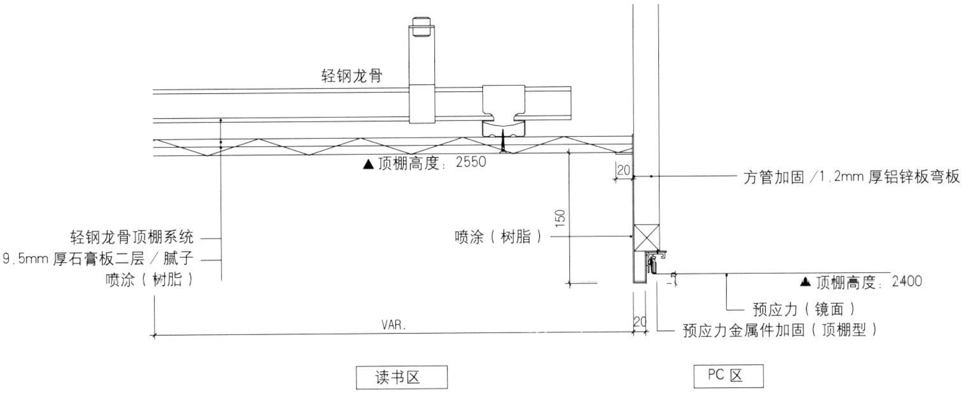

读书区顶棚一剖面图

■ 二层音乐欣赏区——家具

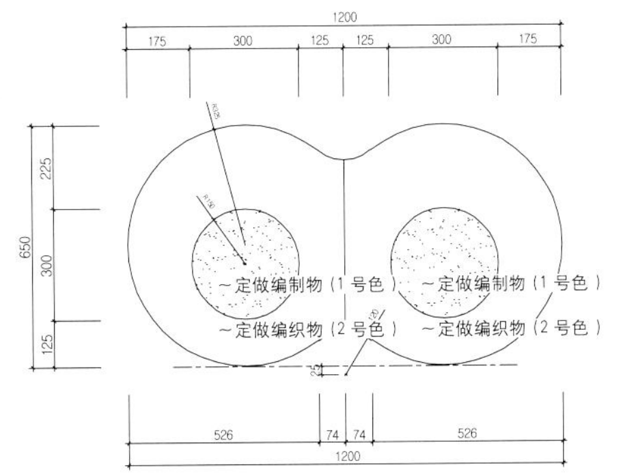

椅子一顶视图

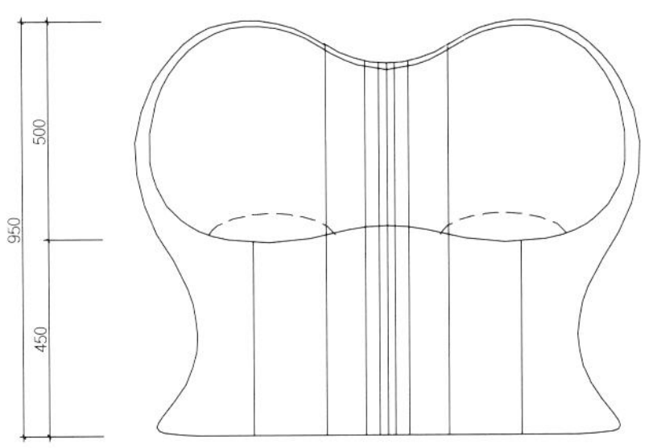

椅子一正视图

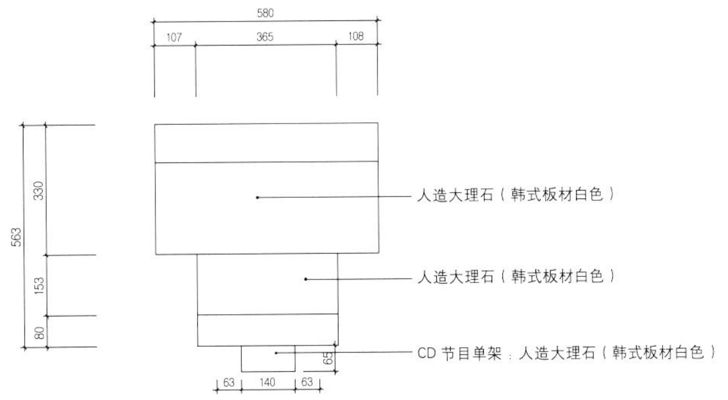

音响一顶视图

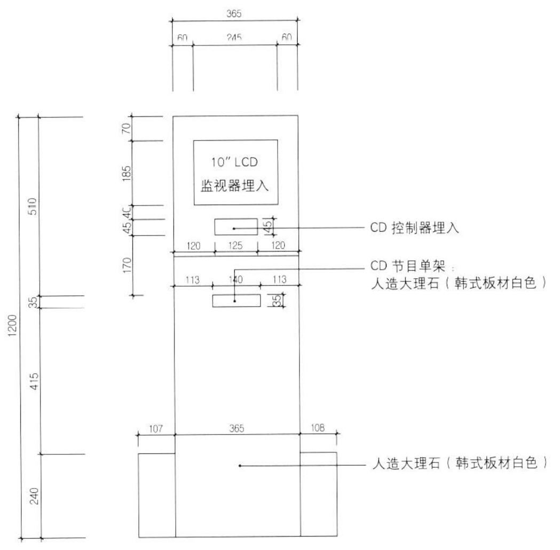

音响一正视图

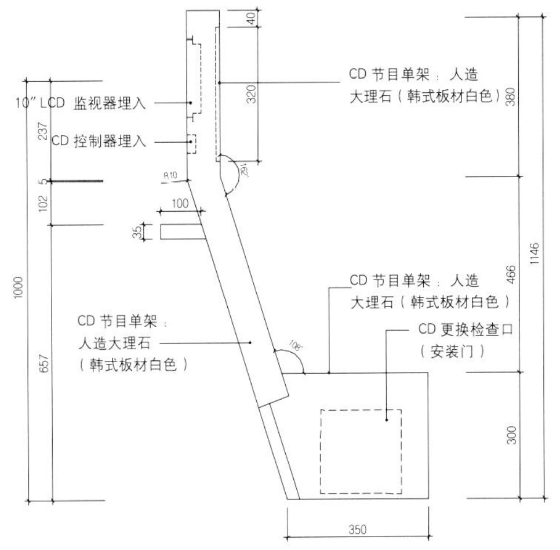

音响一剖面图

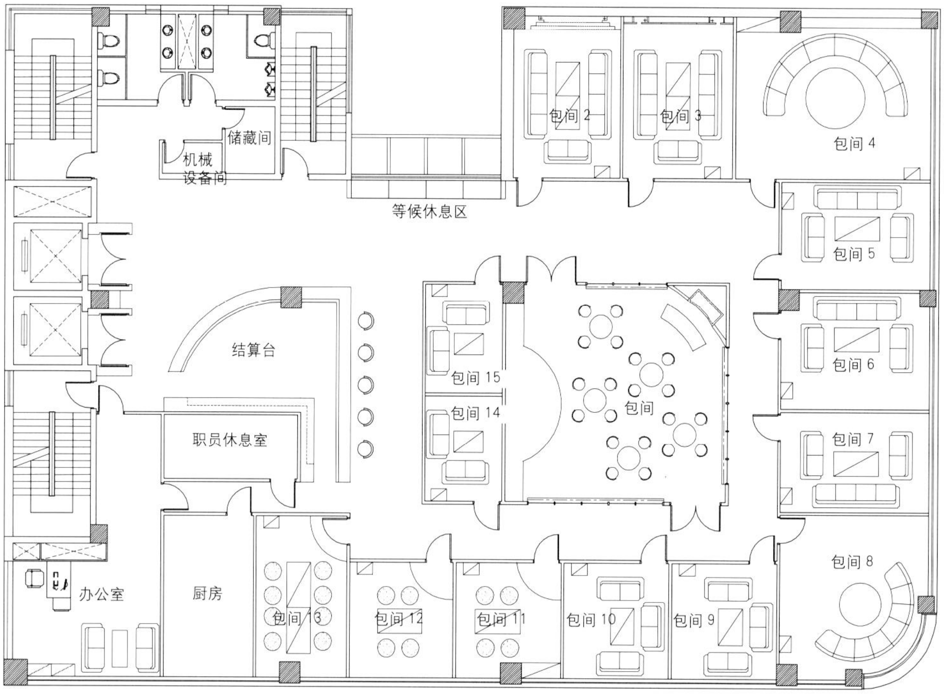
包间 2
包间 3
包间 4
储藏间
机械
设备间
等候休息区
包间 5
包间 6
包间 7
包间 8
结算台
包间 15
包间 14
包间
职员休息室
办公室
厨房
包间 13
包间 12
包间 11
包间 10
包间 9

六层平面图

包间 2
包间 3
包间 4
机械
设备间
储藏间
等候休息区
包间 5
包间 6
包间 7
包间 8
结算台
储藏间
包间 14
包间 13
包间 1
职员休息室
储藏间
包间 16
包间 15
办公室
厨房
包间 12
包间 11
包间 10
包间 9
0 1 3 5 10

五层平面图

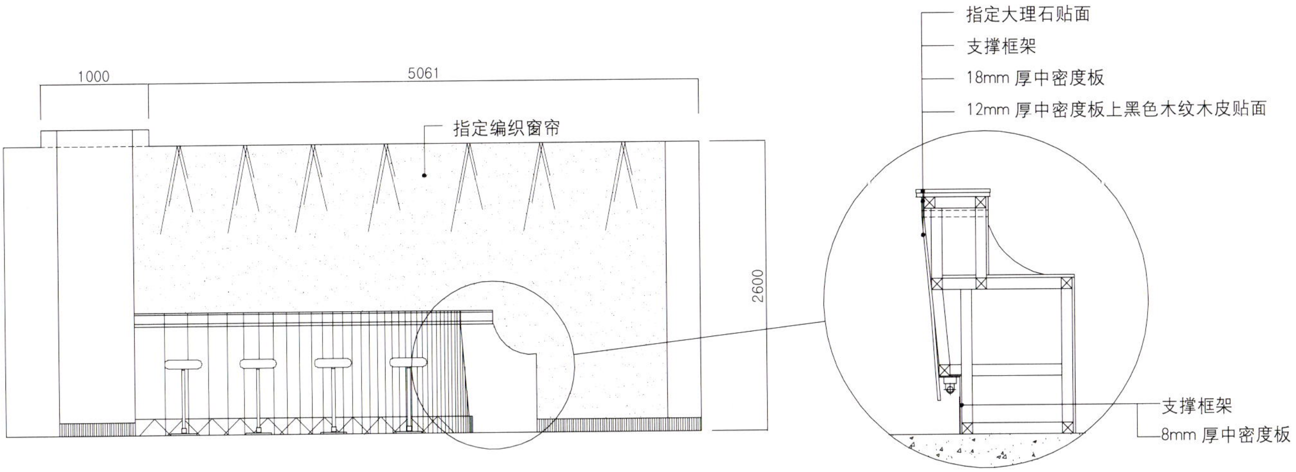

五层吧台结算台一立面图

设计：真空间设计研究所 李明锡（02）517-2051 | **施工图设计：**真空间设计研究所 金宝庆 南铉戊 | **施工：**真空间设计研究所 | **地址：**忠清南道天安市成正洞 | **面积：**1016.4m² | **装修：**地面／釉面砖、地热阳台、雕刻瓷砖 墙面／腻子、韩宣纸墙纸、陶瓷锦砖 顶棚／树脂涂料 | **摄影：**李峰哲

Design: Charm Space Interior Design · Lee, Myung hee I **Execution Design:** Charm Space Interior Design·Kim, Bo kyung / Nam, Hyun uk I **Construction:** Charm Space Interior Design I **Built Area:** 1016.4m² I **Finish:** Floor / Polising Tile, Heated Floor, Deco Tile Wall / Antico Stucco, Korean Paper, Mosiac Tile Ceiling / V.P I **Photograph:** Lee, Bong chul

■ 五层吧台

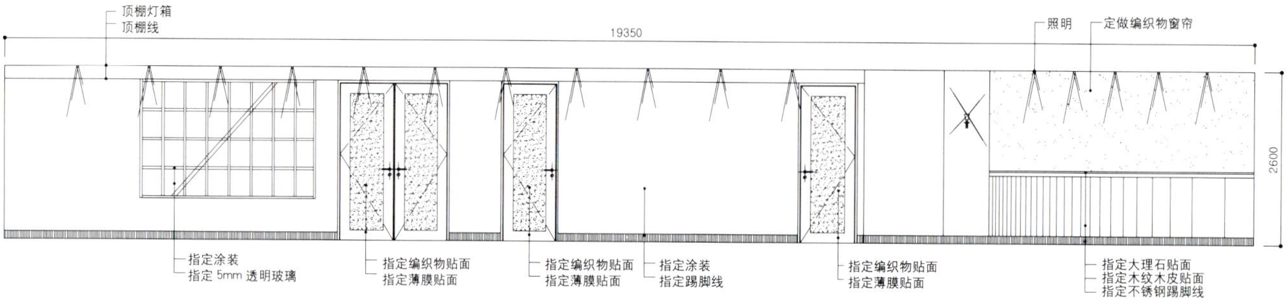

立面图

■ 五层包间1号

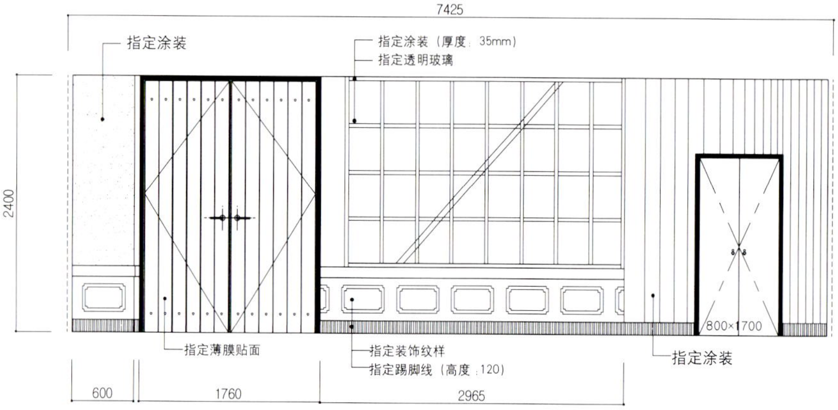

立面图 A

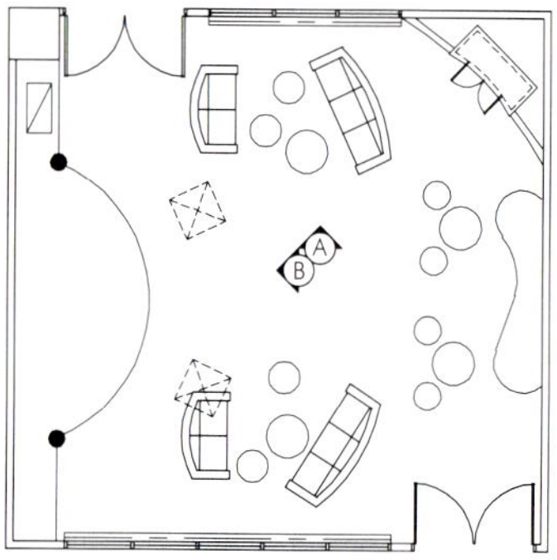

局部平面图

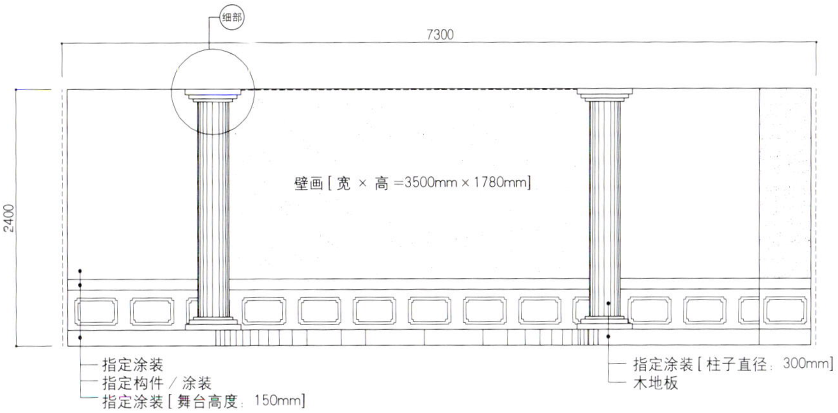
细部
7300
2400
壁画[宽 × 高 =3500mm × 1780mm]
指定涂装
指定构件 / 涂装
指定涂装[舞台高度：150mm]
指定涂装[柱子直径：300mm]
木地板

立面图 B

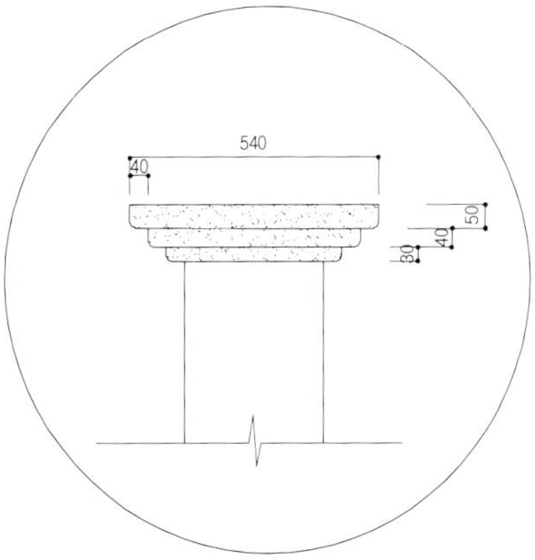
540
40
50
40
30

细部图

现代社会快节奏的生活和工作方式是不是使你感到压力很大？如果这些压力集聚到了一起，又得不到释放，这个世界将会变得怎样？这些现代人会怎么样呢？当前的这种状况下，大声唱歌就是解除压力的好方法。以前的所谓练歌房规模较小，封闭的空间中给人的感觉过于单一。像在火柴盒狭小的空间中一样，对于想提高嗓门唱歌的现代人，每一次都想彻底放松一下，所以就想起了类似电影院音响效果的空间。比起原先的四面实墙，这里宽大的窗户可以对着外边的世界呼喊，沉浸在旧时电影场景中……看着吧台上鸡尾酒表演……这就是李明锡所长在这个设计中想展示的空间感受。她对每一个房间都布置了不同的颜色、不同的题材。把不同的感觉和文化集合在一起，从而创造了新的游戏空间。她现在又开始第二个明星之塔的计划了。希望这个新的明星之塔是缓解现代人伤痛的地方。

■ 五层包间3号

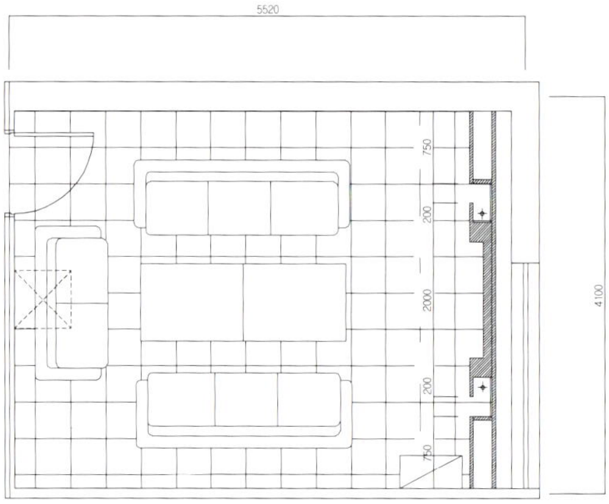

局部平面图

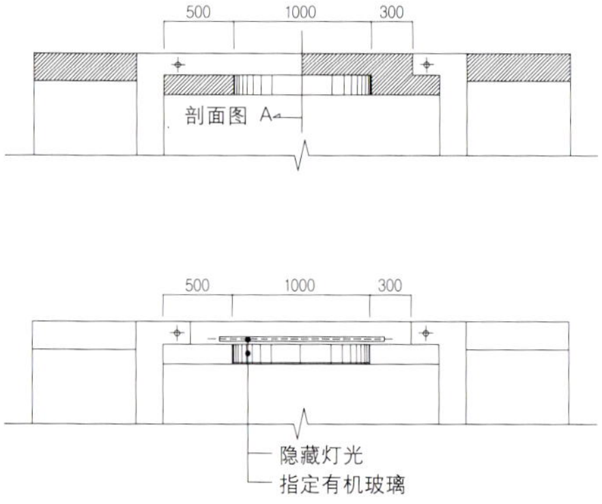

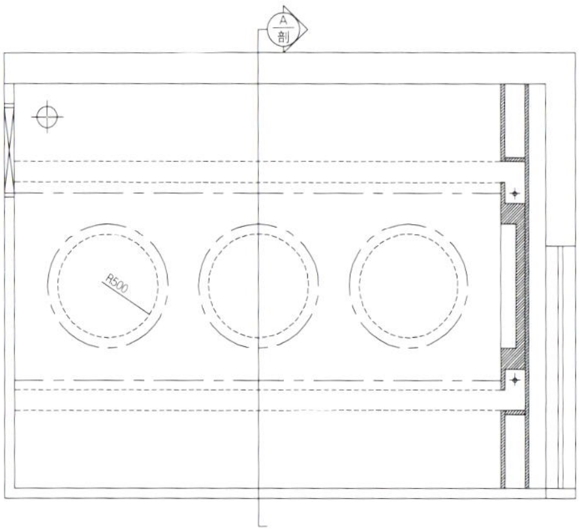

局部顶棚平面图

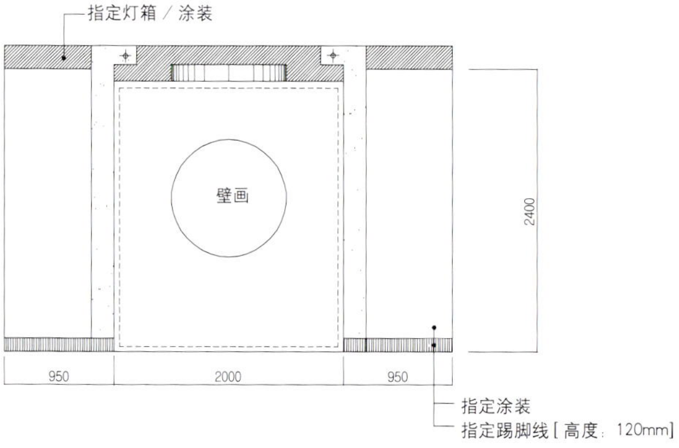

剖面图 A

■ 五层包间4号

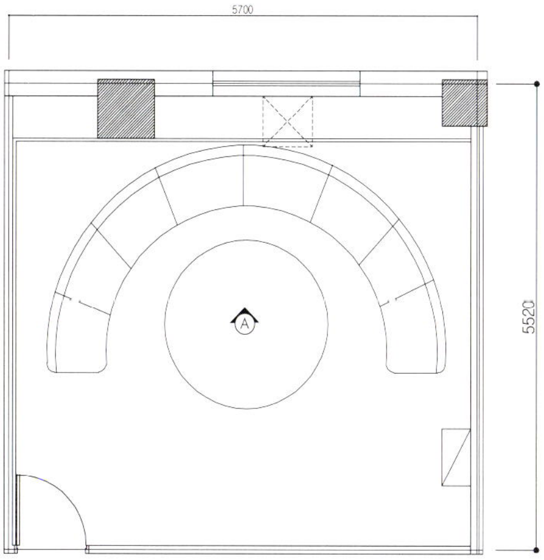

局部平面图

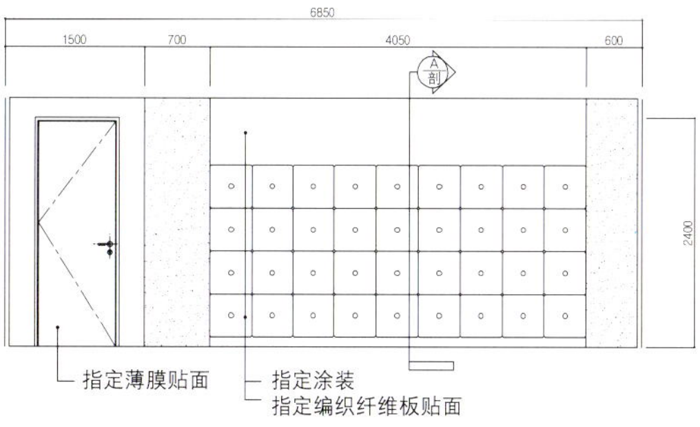

立面图 A

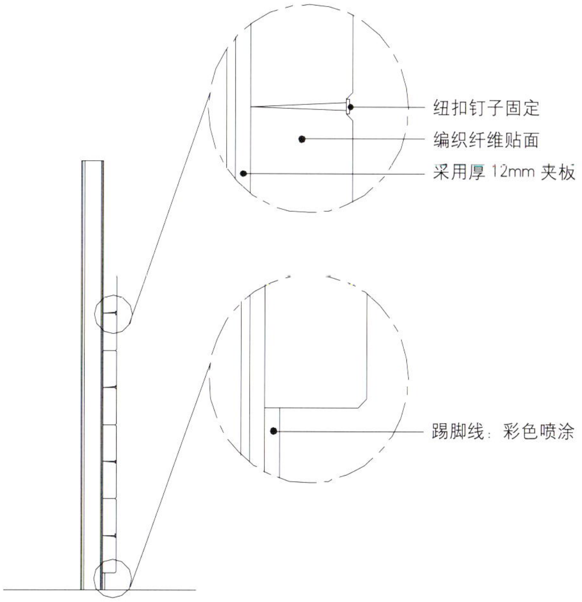

剖面图 A

What would happen to the world if people cannot blow their stress away? What would happen to the people, then? Singing out loud could be one way to live this age. An ordinary singing room has a standardized feel coming from a small, closed space. The designer thought of a cool space like those in the movies for modern people, who sing at the top of their voices in a space small as a matchbox. A space with a large window through which one can shout out to the world, a space that draws people to the old movies. A space that presents a cocktail show in a buoyant bar at least for waiting moments. Those are what director Lee Myung hee intended. She designed each space with different colors and stories. Various senses and cultures gathered to be born again as a new entertainment space. Now she has a plan for the second Star Tower. The designer hopes that it would be a 'tower' that soothes the wounds and weariness of modern people, not a space for the celebrities.

■ 五层包间10号

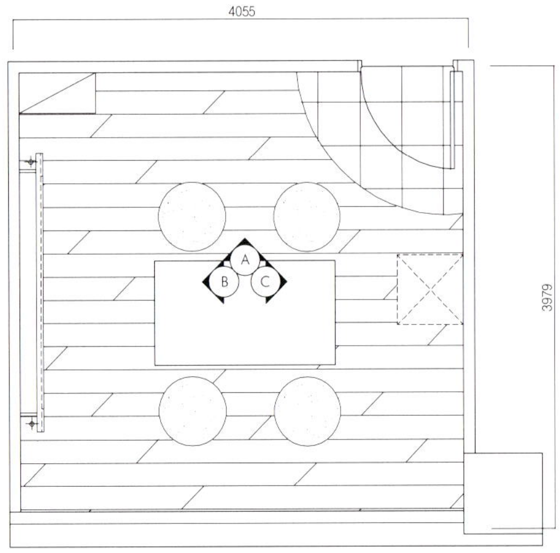

局部平面图

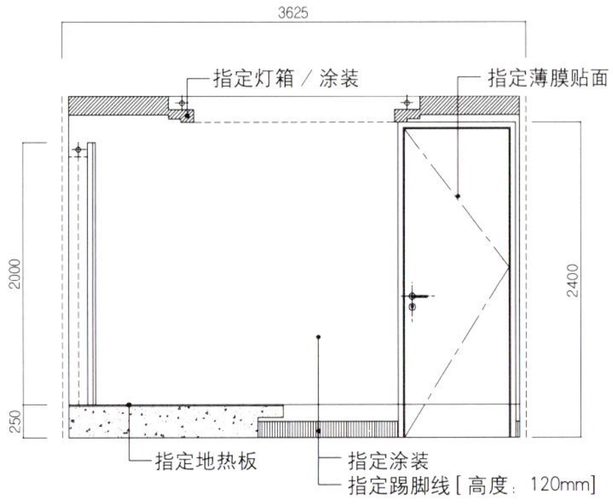

立面图 A

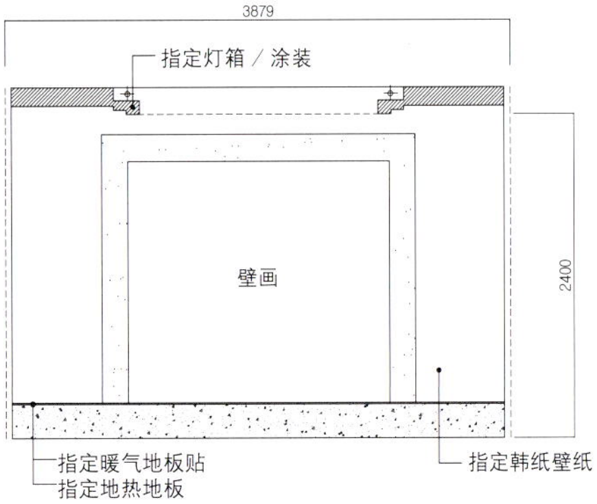

立面图 B

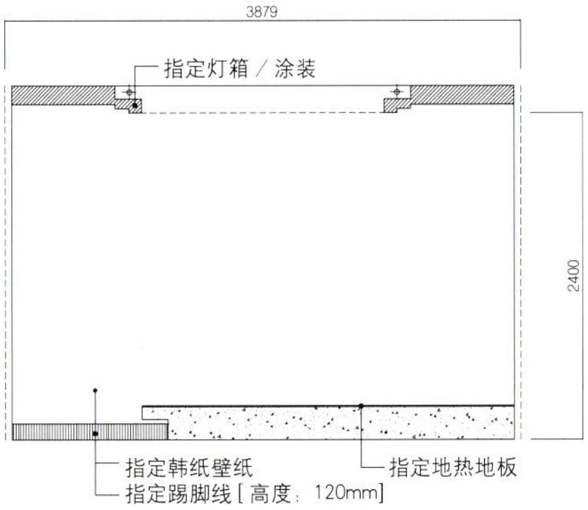

立面图 C

■ 六层包间1号

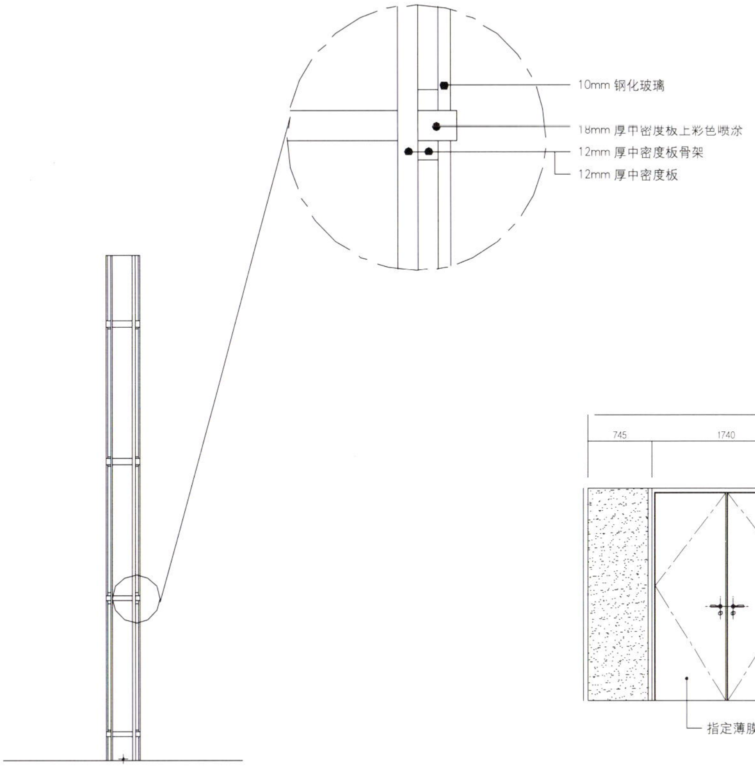

剖面图 A

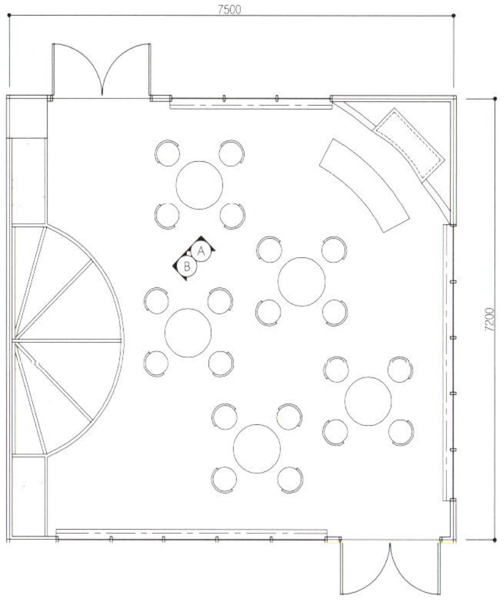

局部平面图

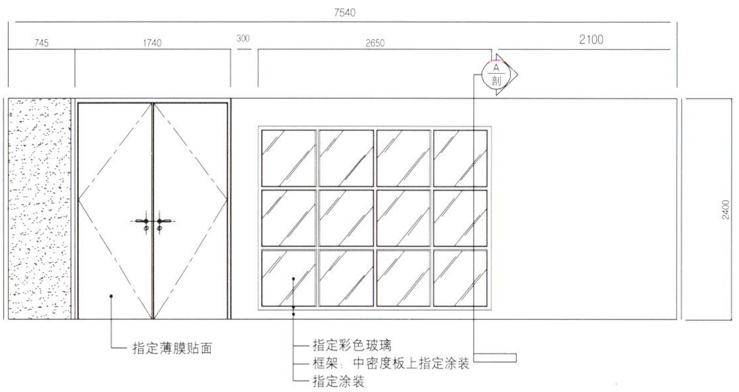

立面图 A

■ 六层包间1号

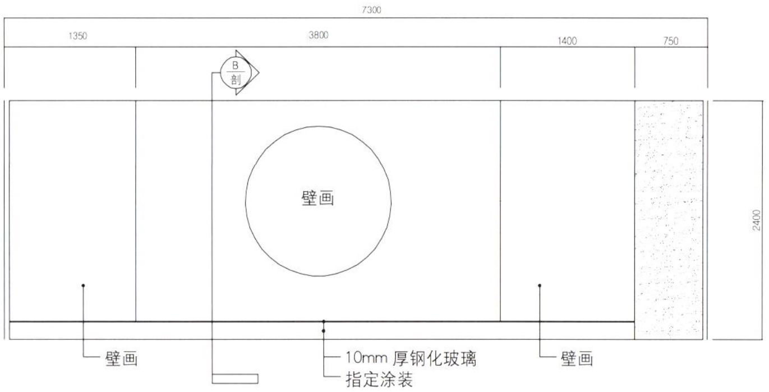

立面图 B

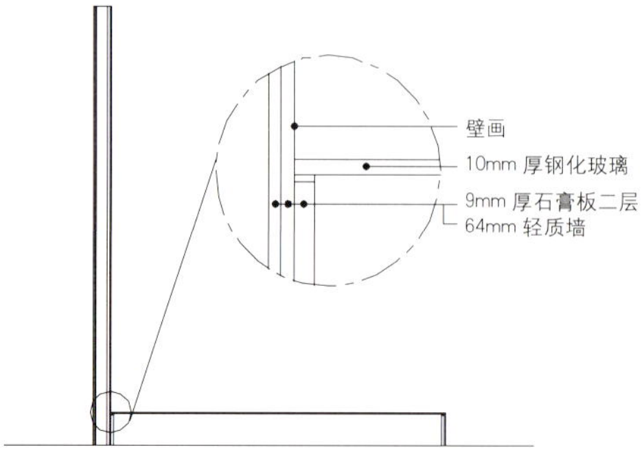

剖面图 B

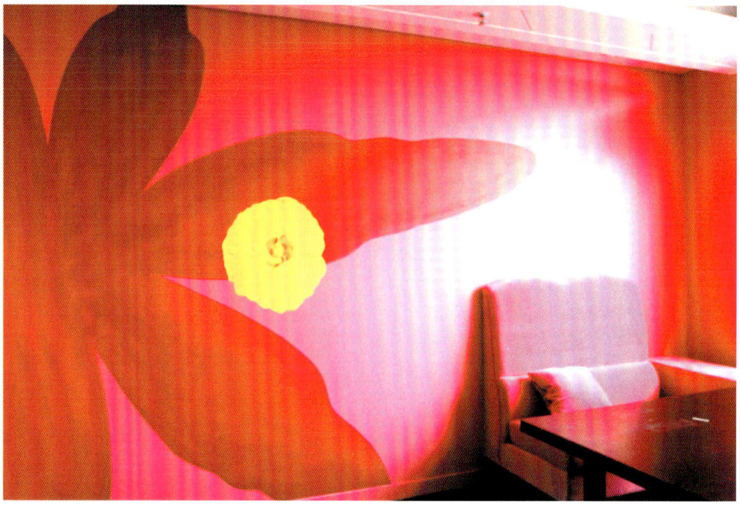

■ 六层包间6号

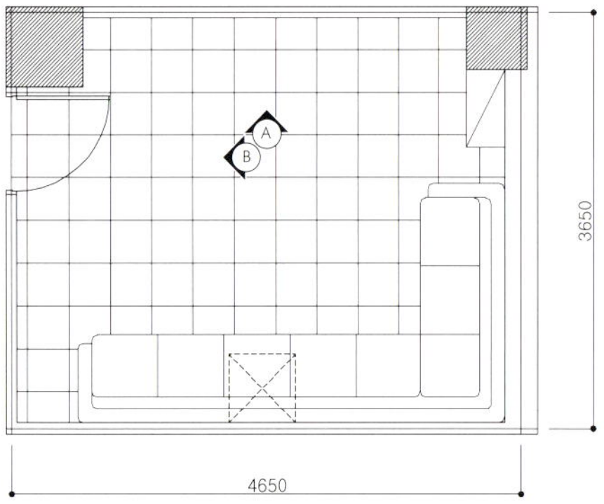

局部平面图

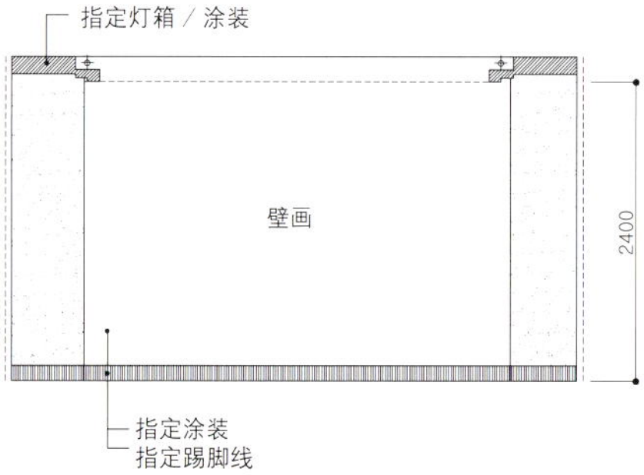

立面图 A

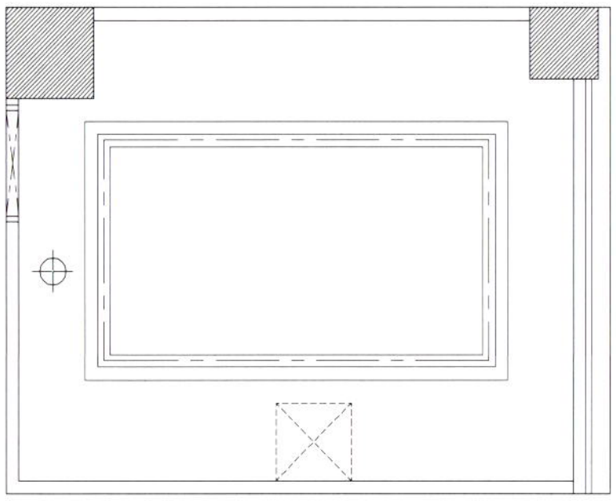

顶棚平面图

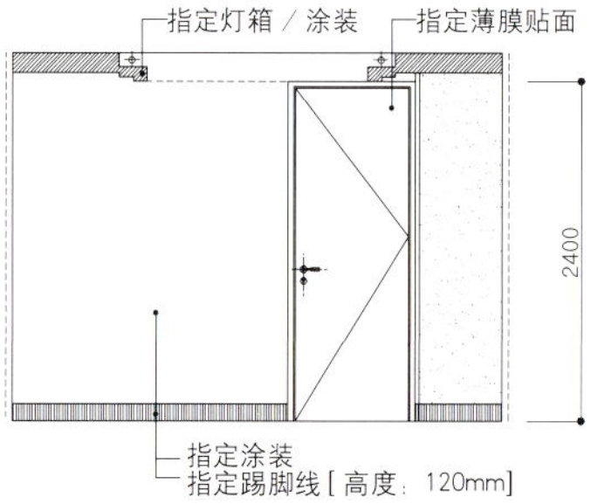

立面图 B

■ 六层吧台结算台

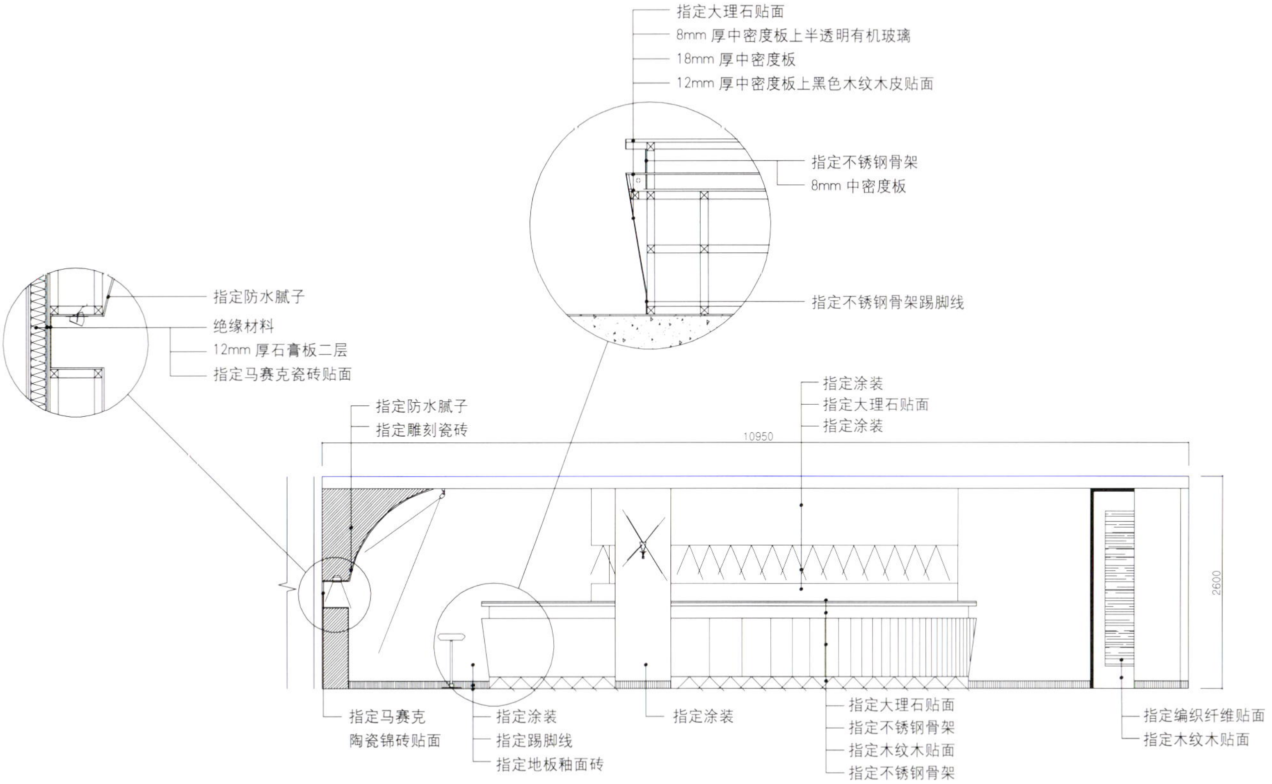

指定大理石贴面
8mm 厚中密度板上半透明有机玻璃
18mm 厚中密度板
12mm 厚中密度板上黑色木纹木皮贴面
指定不锈钢骨架
8mm 中密度板
指定不锈钢骨架踢脚线
指定防水腻子
绝缘材料
12mm 厚石膏板二层
指定马赛克瓷砖贴面
指定防水腻子
指定雕刻瓷砖
指定涂装
指定大理石贴面
指定涂装
10950
2600
指定马赛克
陶瓷锦砖贴面
指定涂装
指定踢脚线
指定地板釉面砖
指定涂装
指定大理石贴面
指定不锈钢骨架
指定木纹木贴面
指定不锈钢骨架
指定编织纤维贴面
指定木纹木贴面

立面图

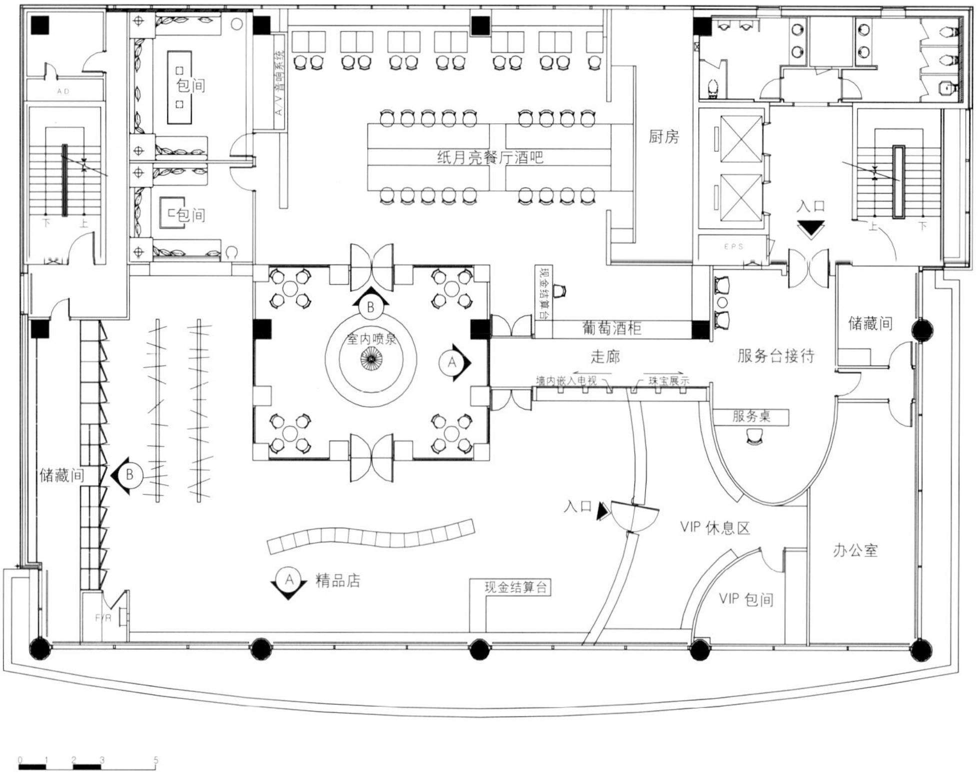

平面图

最近，在首尔市论岘洞和平大厦的三层出现了一个特别的场所——罗拉会员店。该店从入口开始就充满了令人期待的气氛。这里陈列了很多名牌产品，有名牌商品柜台、珠宝柜台以及酒吧等，设计师试图将功能各不相同的这三种空间揉和在一起，形成一个相互协调的大空间，从而创造一种高雅、中庸、时尚和舒适的氛围。

罗拉会员店的成员由罗拉连锁店及纸月亮餐厅酒吧组成。从入口处开始的柔和粉红色透露出女性的气质。沿着走廊进入室内是围绕喷泉的大厅，大厅的右侧是纸月亮酒吧，左侧是品牌店。从气氛上来看，明亮的地方是品牌店，稍暗的地方是纸月亮酒吧。站在此处究竟应该去哪里确实费脑筋。另外，两处各具独特个性的商店被新古典主义喷泉和木质门所划定的中间地带自然地连接分割开了。从而有效地化解了两者之间的冲突。

罗拉会员店是按照现代高雅的标准装修的。多种名牌商品集聚一堂，简洁中某些亮点提高了层次。波浪形的展示柜与顶棚的造型物，采用原色调的固定柜台等将琳琅满目的商品展示在你面前。

后面浅蓝色的纸月亮采用了现代的素材，强调线型的酒吧呈十字形排列。蓝色霓虹灯的光线和黑镜桌子是很协调的感性空间。另外，通过黑镜处理的吧台和镜子能够看到对方的侧面轮廓以及喜怒哀乐，极富视觉趣味。

因此说从原先的框架中创造出新的空间感受是非常令人兴奋的。

设计：Kutec 设计公司 郑圭太（02）2268–4960 | **施工图设计**：Kutec 设计公司 金孝锡、孙秉国、徐艺荣 | **施工**：Kutec 设计公司 | **地址**：首尔市江南区新沙洞 629–15 和平汽车江南专卖店三层 | **面积**：765m² | **装修**：地面 / 釉面砖、钢化地板、雕刻瓷砖 墙面 / 耐火砖、漆喷涂、编织纤维布、黑镜面 顶棚 / 树脂喷涂

Design: Design Kutec · Chung, Kute I **Execution Design:** Design Kutec · Kim, So hui / Son, Byeong guk / Seo, Ye yeong I **Construction:** Design Kutec I **Built Area:** 765m² I **Finish:** Floor / Polishing Tile, Compressed Floor, Deco Tile Wall / Terra-Cotta Coating, Lacquer, Fabric, Black Mirror Ceiling / V.P

■ 服务台

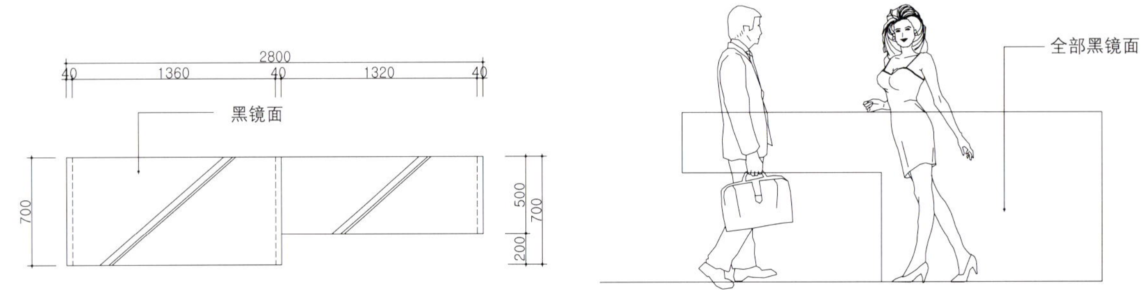

顶部和前部视图

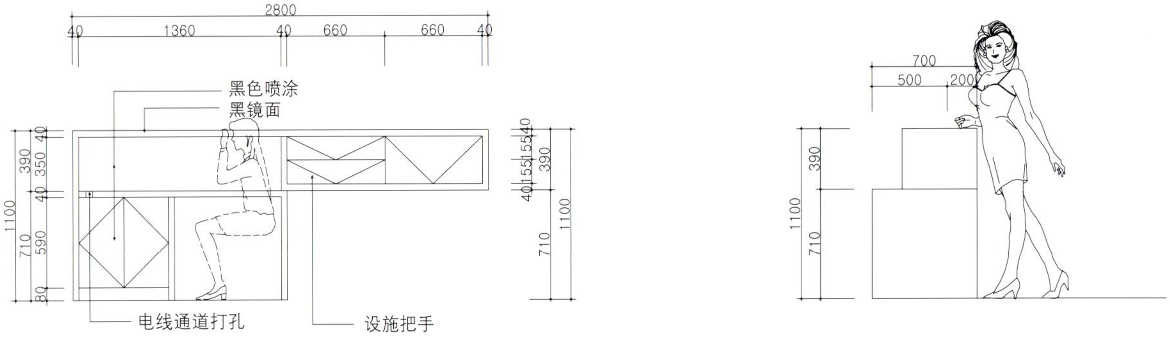

后视及侧面图

■ 室内喷泉

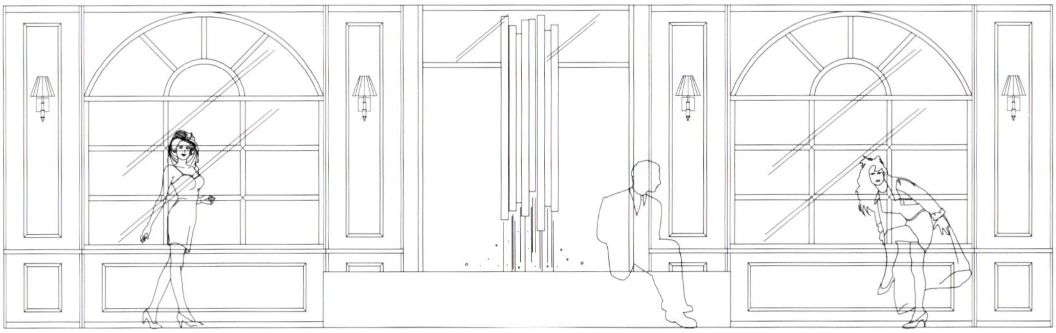

装饰画

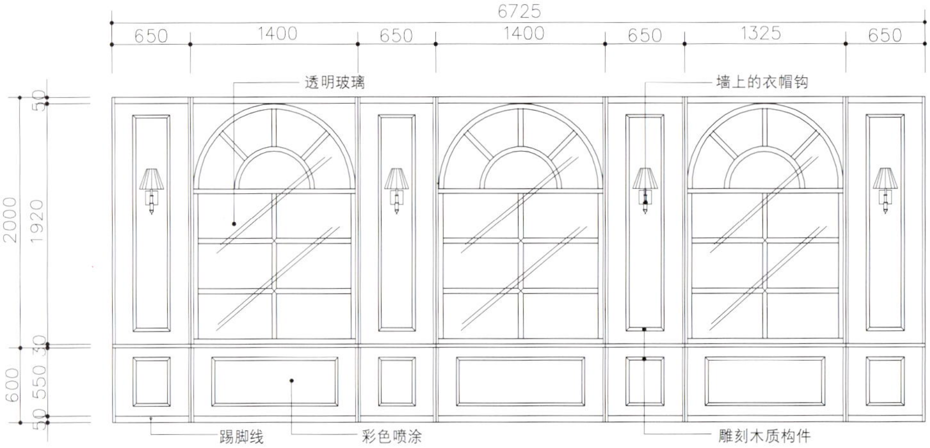

立面图 A

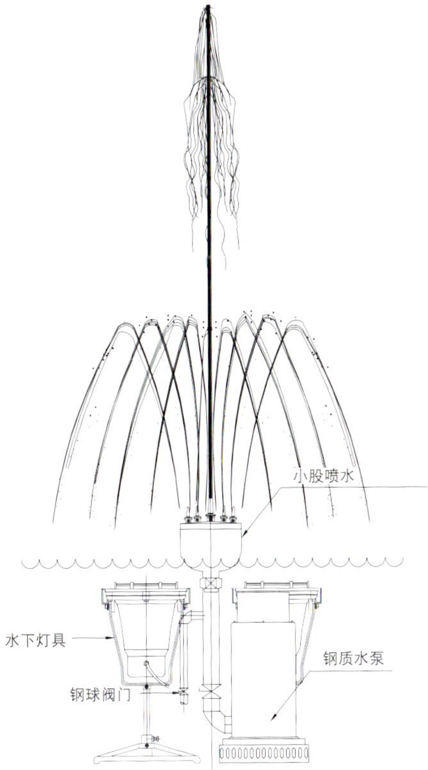

剖面图

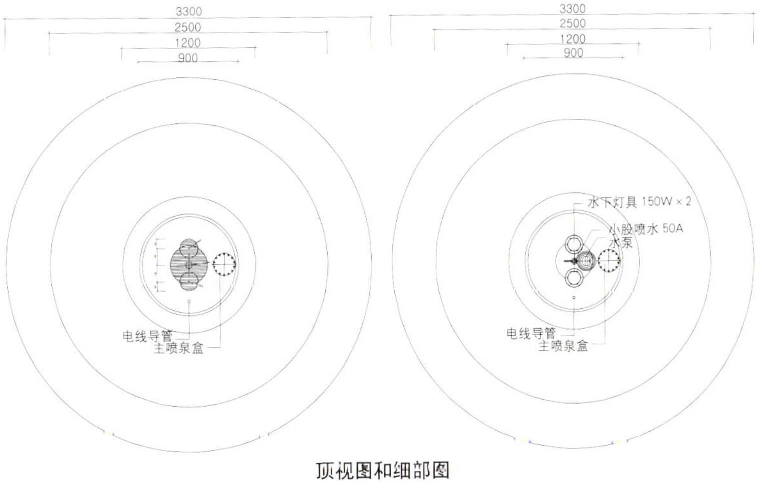

顶视图和细部图

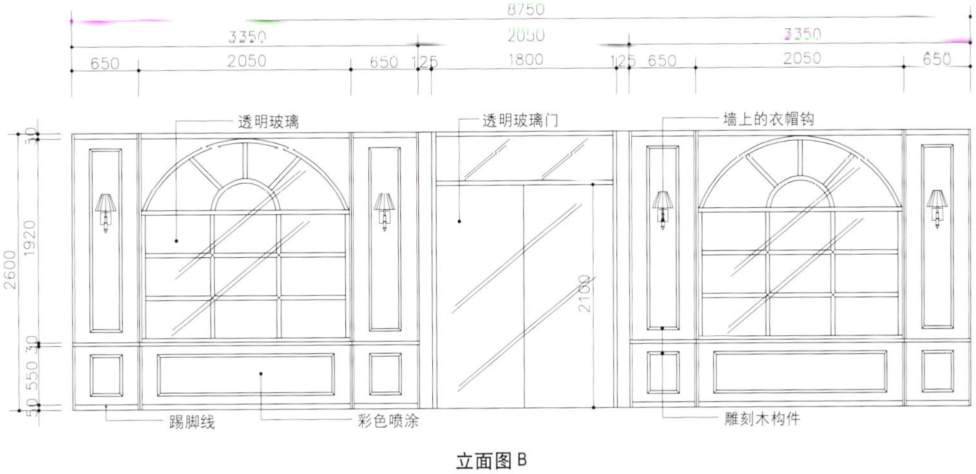

立面图 B

A quite unite space has showed up on the 3rd floor of Pyeonghwa Building, Nonhyeon-dong. The expectation begins from the entrance of LOA Members. It is a coexisting space of a bar and a jewelry/brand shop stocked up with various products of luxury brands. Three spaces with different functions have met. Thus it was important to harmonize them into a unified space. The designer attempted to hold moderation, elegance, trend and comfort all together.

LOA Members consists of multi-shop LOA and restaurant/bar Papermoon. Mild pink of the entrance gives out femininity. A lounge with a fountain at the end of a hallway has the bar on the bright right side and LOA on the dark left side, leaving people with a difficult choice where to go first. The central area composed of a neo-classical fountain and an antique wood door naturally connects and divides the individual atmospheres of the two shops, smoothing the possible awkwardness.

LOA radiates a modern, elegant feel. Because of the product diversity, simple and accentuating elements were added. A wave-shaped display case, a plastic structure of the ceiling and a raw color built-in closet fill the space and stand out as to the products.

In the bluish Papaermoon is a cross-shaped bar that showing of its material and lines of modern feel. It is a sensuous space where the blue neon light and a black mirror table match to each other. Black mirrored bar and mirror increases the visual amusement through the reflected looks or silhouettes of one's companion.

It was an exquisite experience to see a space outside the existing frame.

■ 精品店

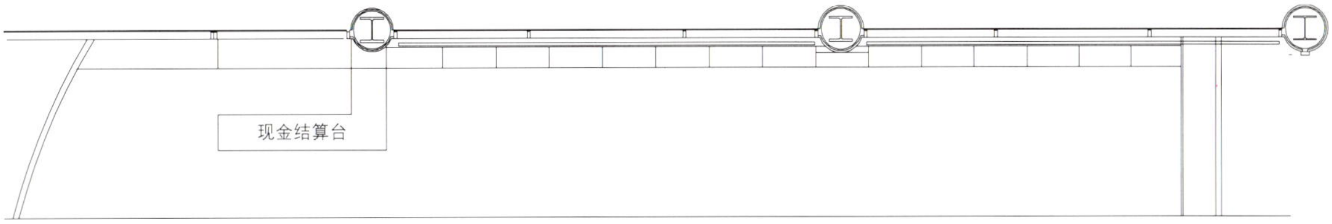

壁柜—地板平面图 A

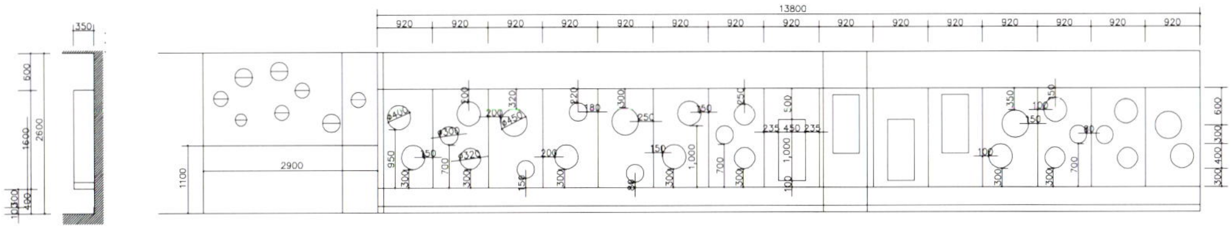

立面图

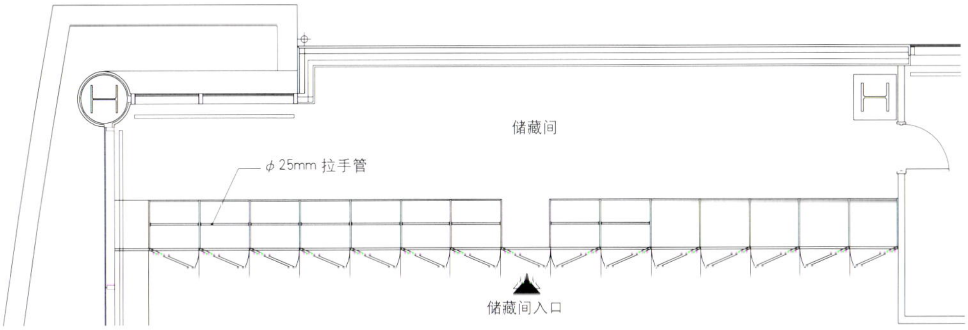

壁柜平面图 B

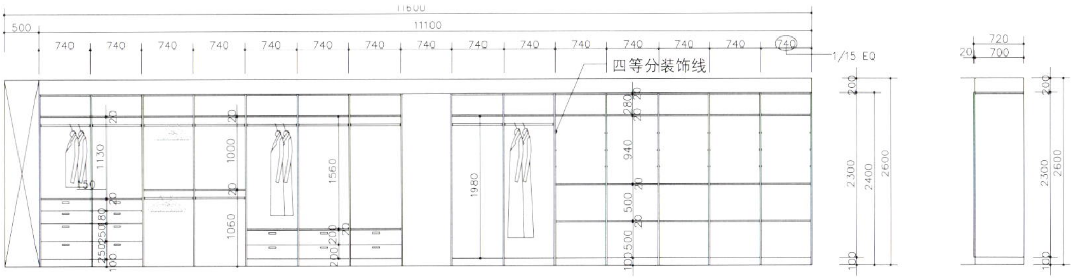

立面图

Cheongjeong Daryewon, Pohang

浦项清静茶礼园

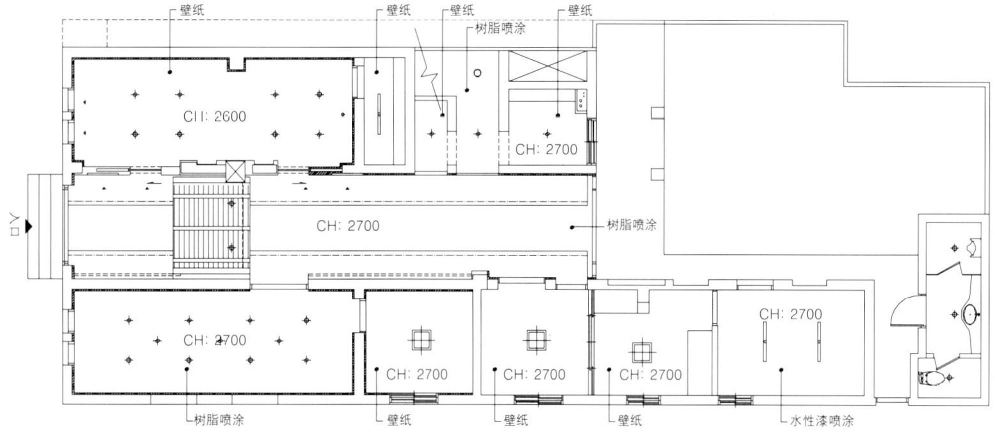

一层顶棚平面图

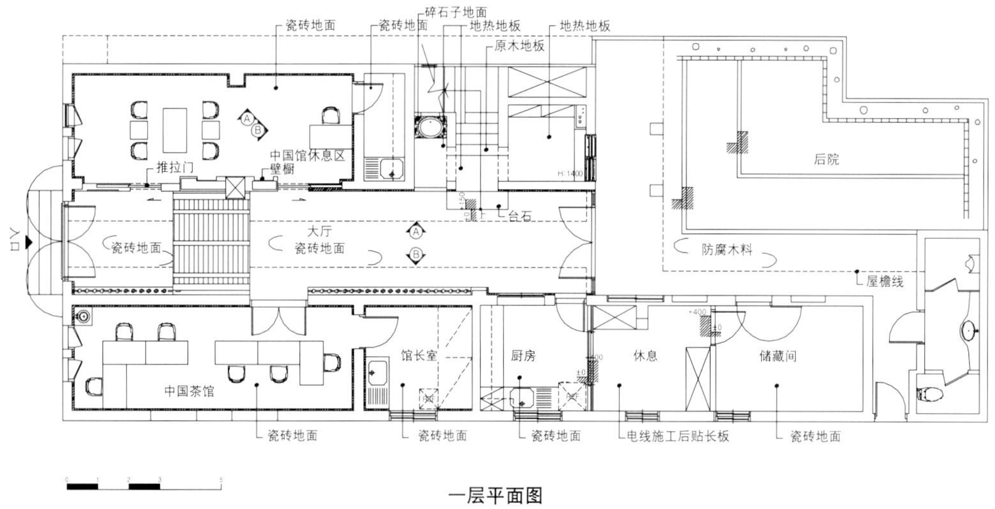

一层平面图

建筑业主非常富有社会奉献精神，而这栋二层茶道会馆就是业主捐献出来的。茶礼园的设计出发点是为了表现韩、中、日三国的传统茶文化。入口处装饰的瓦片是东亚三国均具有的特有建筑构件，这些构建均是按照规格特别烧制的。坐在高高的椅子上饮茶是中国人的习惯，因此中国茶室位于一层，其对面是茶叶展览休息室。脱掉鞋子在日式洗手盆洗过手以后就可以登上二层，在二层，日本茶室和韩国茶室相对而设，这里是要求客人坐在地板上饮茶的。茶室名称根据业主的号命名为茶慧室，它位于内房，即家中女性的房间，日本茶室位于外房，即家中男性的房间，取名为高恩室。日本茶室尽量再现传统的日本茶文化，东京室旁边的柱子是特意从山上砍来并且用镰刀细心地加工过的。另外，会馆经常举办茶道讲座，专门为来此学习茶道的普通人及外国人举办培训班。也为他们的驻留配备了洗澡间等便利设施。

因为，该项目是公司第一次运作，虽然期间遇到了很多困难也有一些遗憾，但是，仍然希望都市人能摆脱城市中心区生活和工作的烦恼，到这里来体会放松的快乐。

The 2nd story tea ceremony center, holding a public service spirit of its owner, focuses on reproduction of Korean/Chinese/Japanese traditional tearooms. Roof tiles at an inner entrance were expressed in an interesting way, perfectly representing Orientalism, being a common architectural element of three countries. 1st floor has a Chinese tearoom and on the opposite a waiting room displaying exhibits. After taking off shoes and washing hands in a Japanese washbowl, one can see Japanese and Korean tearooms on the 2nd floor. As for Korean ones, Dahye room, titled after the owner's pen name, indicates femininity and Goeun room masculinity. Japanese one was a successful reproduction of original. A pillar beside tokonoma is a work by hands and a sickle. It offers shower/convenience facilities to public and foreign people who wish to learn the tea ceremony.

The project was hard being our first project. It still has a room of improvement, but I hope people would have a wonderful time to find this fairyland in a city.

概念设计：东洋大学室内设计学院 裴俊玄教授 | **设计**：（株）合伙人设计公司 吴郑玄、李恩灵、宋智慧 (02) 568—4749 | **施工**：（株）合伙人设计公司 才源羽、蒋圭浩、金尚列 | 绘图：朴章圭 | **地址**：庆尚北道浦项市 | 面积：240m² | **装修**：地面 / 树脂喷涂、壁纸墙面 / 漆、韩纸、编织壁纸、瓦片顶棚 / 榻榻米、陶瓷瓷砖、豆油长板、大厅地热地板 | **摄影**：李峰哲

Basic Design: Bae, Jun hyeon Prof of Dept.of Interior Design, Dongyang Univ. I **Design:** Design Partners · Oh, Jeong hyeon / Lee, Eun ryeong / Song, Ji hye I **Construction:** Design Partners · Chea, Won woo / Gang, Gwi ho / Kim, Sang yeol I **Graphic:** Park, Jang gyu I Built Area: 240m² I **Finish:** Floor / V.P, Wallcovering Wall / Lacquer, Korean Paper Wallcovering, Twine Wallpaper, Roofing Tile Ceiling / Tatami Mat, Ceramic Tile, Soybean Oil Floor, Hall Heated Floor I **Photograph:** Lee, Bong cheol

■ 一层休息室

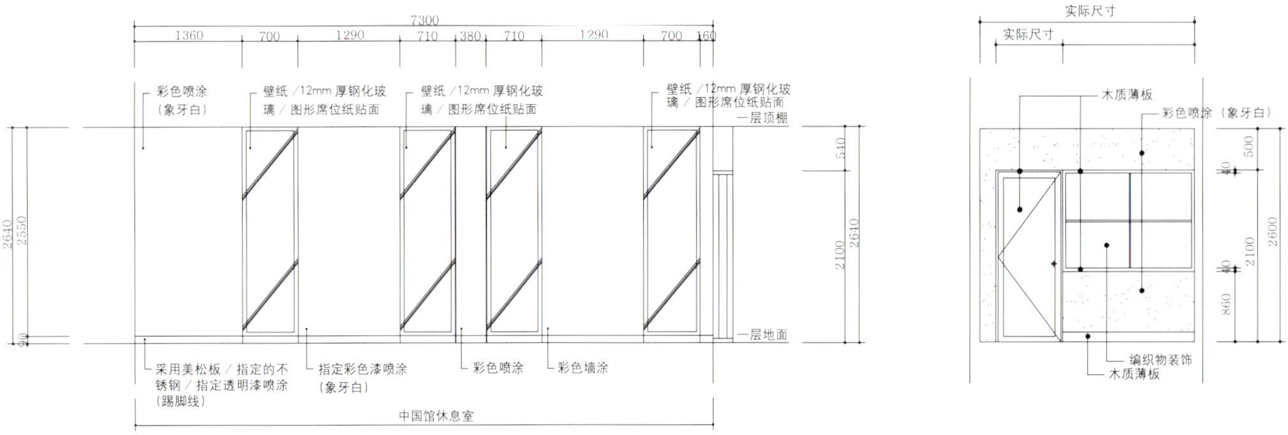

立面图 A、B

■ 一层大厅

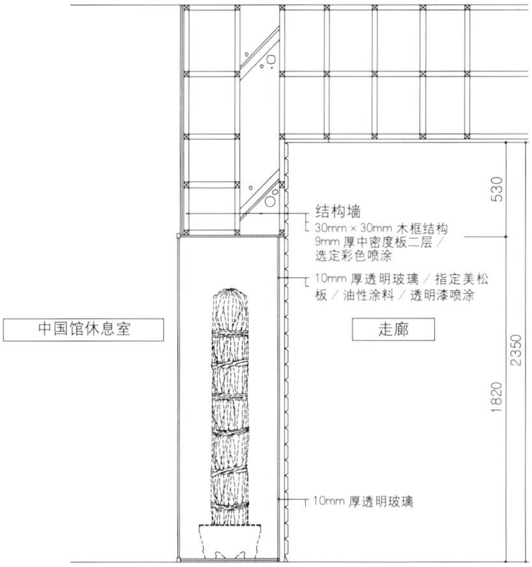

剖面图 A

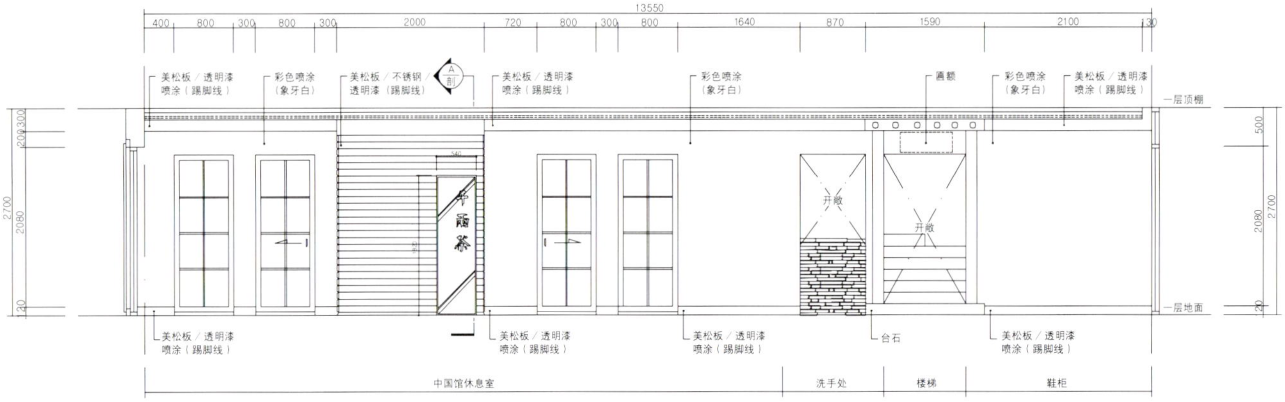

立面图 A

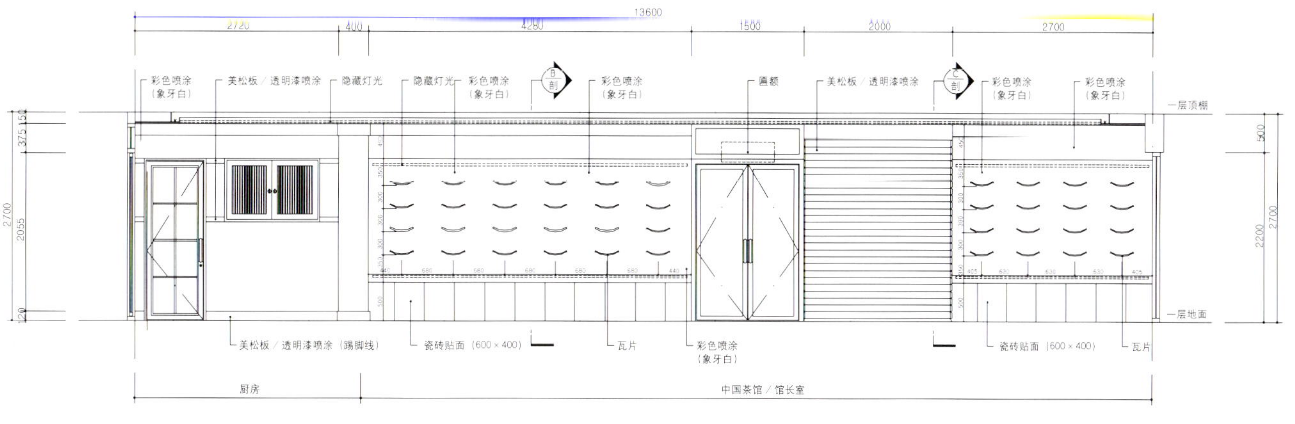
13600
2720
400
4280
1500
2000
2700
彩色喷涂
（象牙白）
美松板／透明漆喷涂
隐藏灯光
隐藏灯光
彩色喷涂
（象牙白）
B
剖
彩色喷涂
（象牙白）
匾额
美松板／透明漆喷涂
C
剖
彩色喷涂
（象牙白）
彩色喷涂
（象牙白）
一层顶棚
一层地面
美松板／透明漆喷涂（踢脚线）
瓷砖贴面（600×400）
瓦片
彩色喷涂
（象牙白）
瓷砖贴面（600×400）
瓦片
厨房
中国茶馆／馆长室

立面图 B（单位：mm）

■ 一层休息间

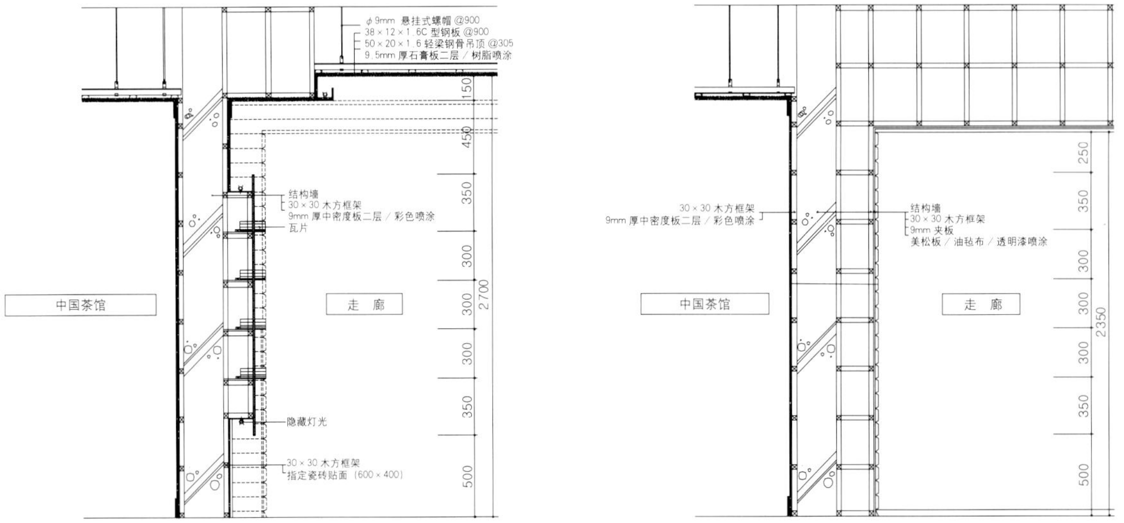

剖面图 B、C（单位：mm）

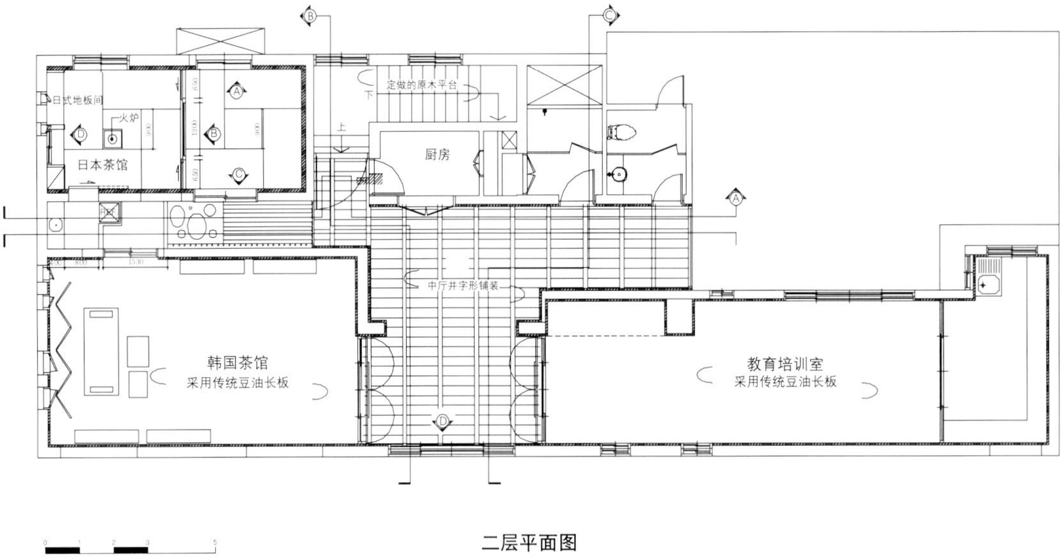

二层平面图

■ 二层中厅

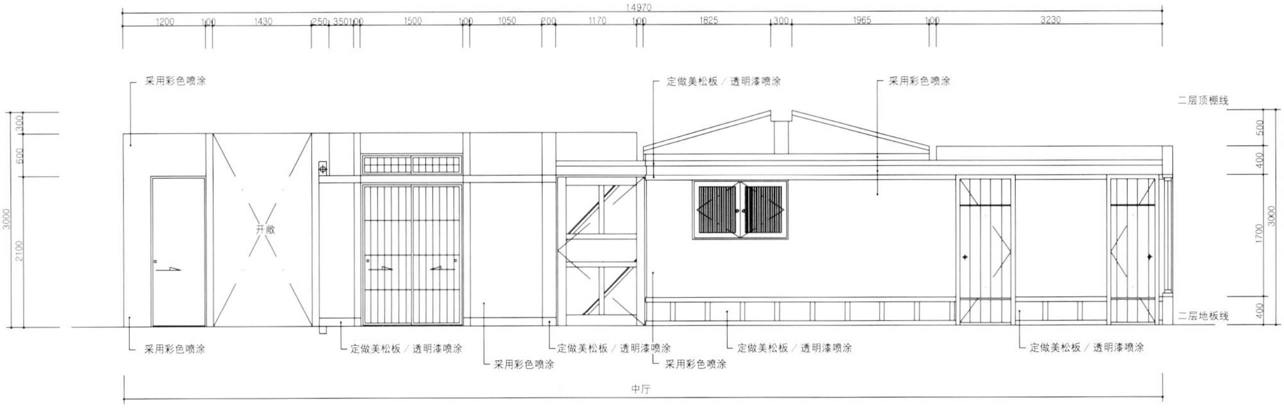

立面图 A

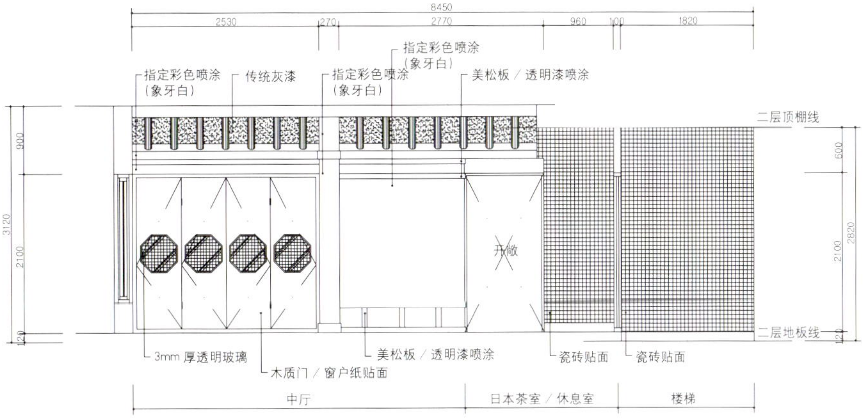
8450
2530
270
2770
960
100
1820
指定彩色喷涂
(象牙白)
传统灰漆
指定彩色喷涂
(象牙白)
指定彩色喷涂
(象牙白)
美松板/透明漆喷涂
二层顶棚线
开敞
二层地板线
3mm 厚透明玻璃
木质门/窗户纸贴面
美松板/透明漆喷涂
瓷砖贴面
瓷砖贴面
中厅
日本茶室/休息室
楼梯

立面图 B

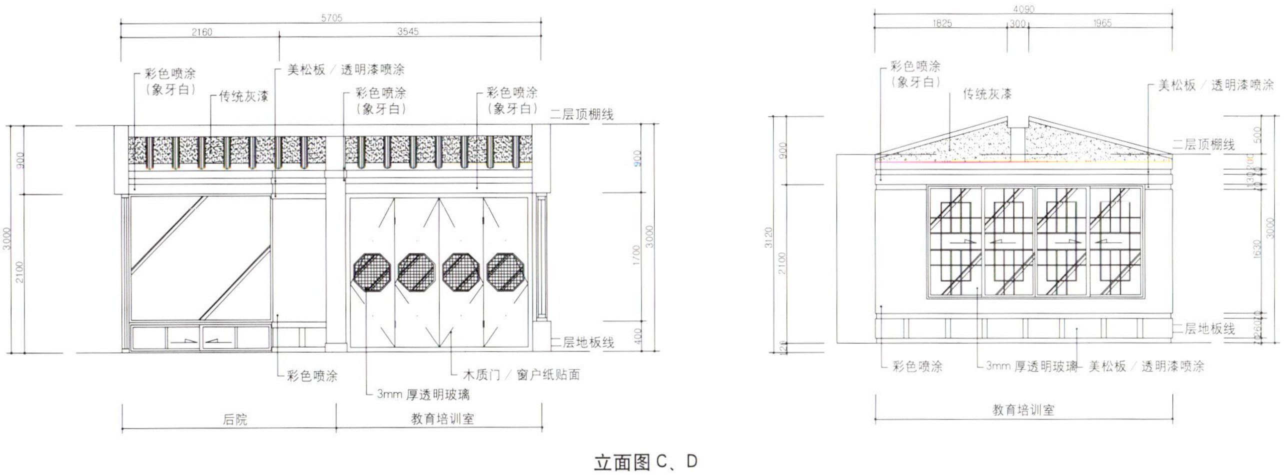
5705
2160
3545
彩色喷涂
(象牙白)
传统灰漆
美松板/透明漆喷涂
彩色喷涂
(象牙白)
彩色喷涂
(象牙白)
二层顶棚线
二层地板线
彩色喷涂
木质门/窗户纸贴面
3mm 厚透明玻璃
后院
教育培训室
4090
1825
300
1965
彩色喷涂
(象牙白)
传统灰漆
美松板/透明漆喷涂
二层顶棚线
二层地板线
彩色喷涂
3mm 厚透明玻璃
美松板/透明漆喷涂
教育培训室

立面图 C、D

■ 二层日本茶馆

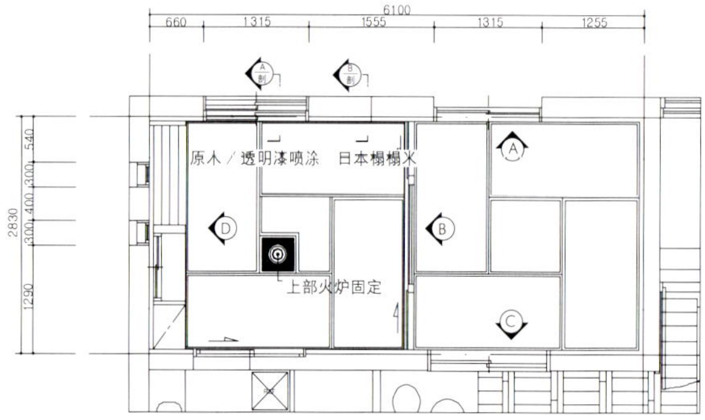

平面图

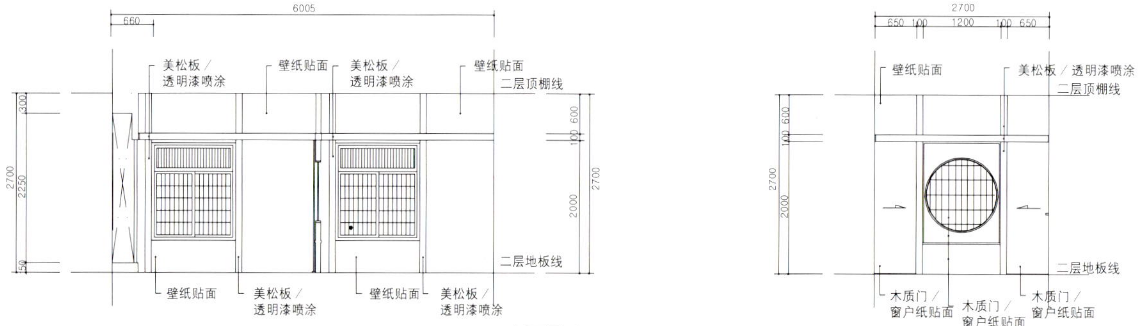

立面图 A、B

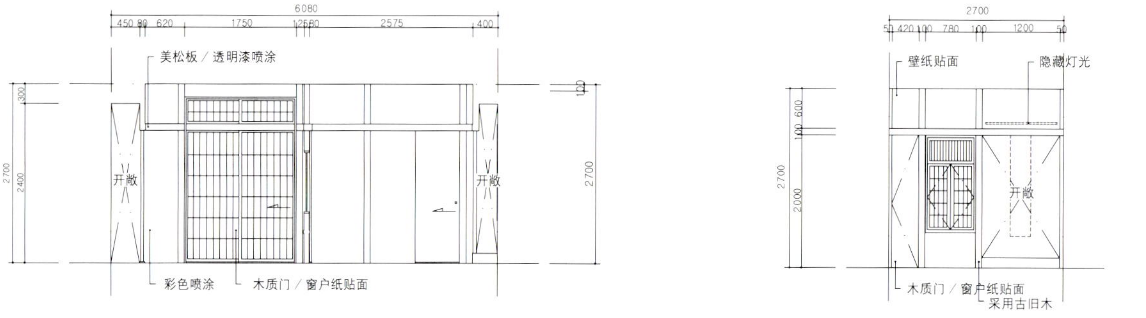

立面图 C、D

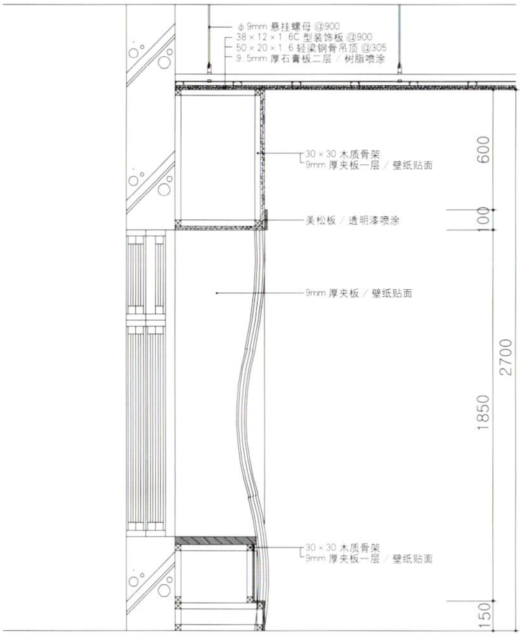

剖面图 A（单位：mm）

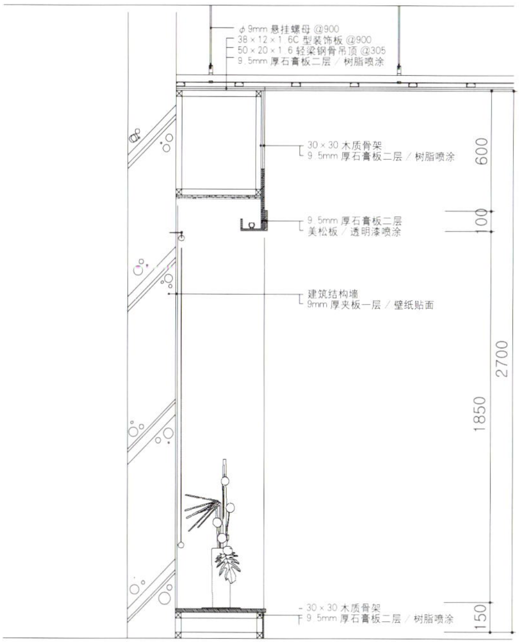

剖面图 B（单位：mm）

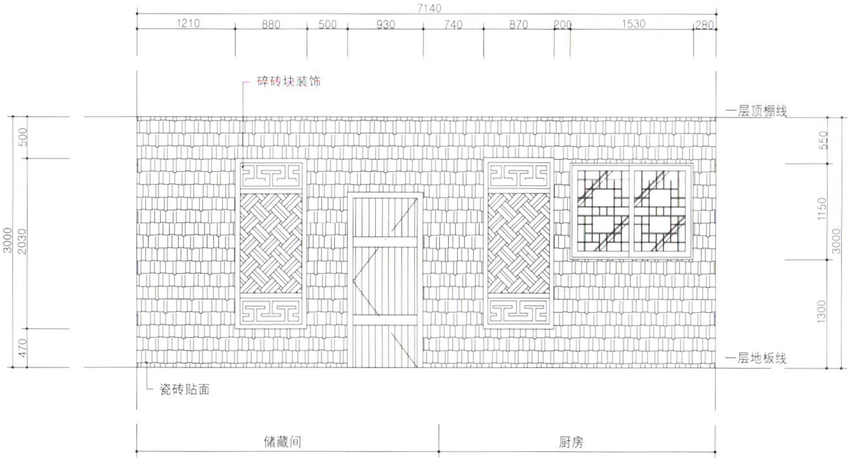

外部一立面图

Boda

包嗒饰品店

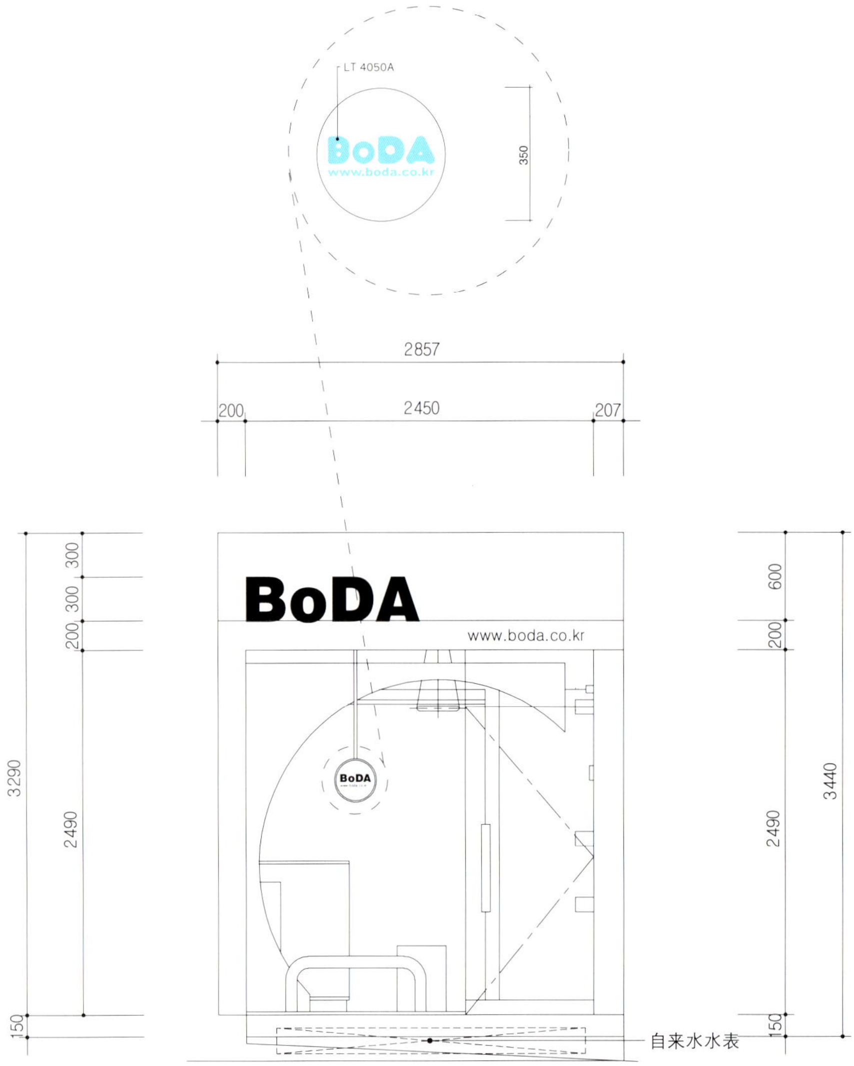

正面图

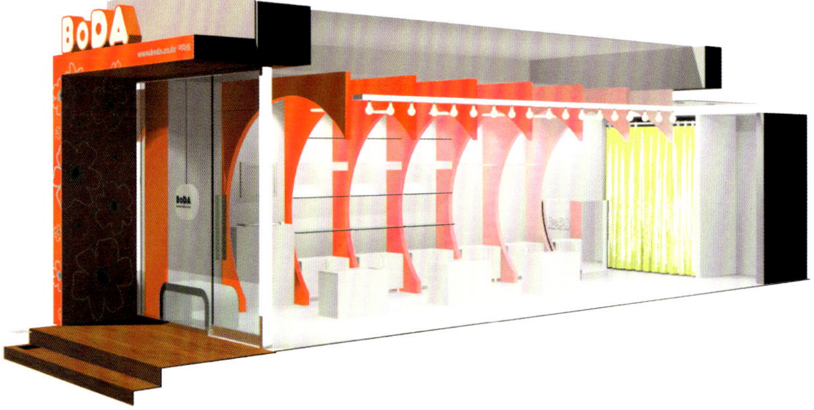

效果图

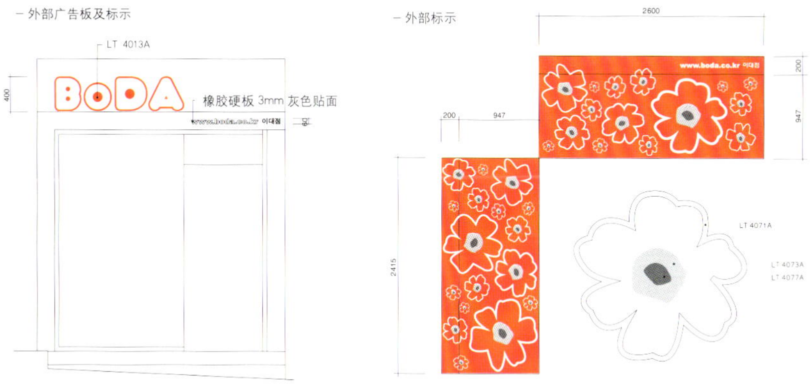

铭牌图

企划设计：朋友设计公司 杨振荣 | **施工图设计：**朋友设计公司 张荣权 （02） 777–9914 | **地址：**首尔市西大门区大玄洞 | **施工：**朋友设计公司 | **面积：**25.82m² | **装修：**地面 / 釉面砖 墙壁 / 彩色喷涂 顶棚 / 彩色喷涂

Planning Design: Friend's Design · Yang, Jin young I **Execution Design:** Friend's Design · Jang, Young gwon I **Construction:** Friend's Design I **Built Area:** 25.82m² I **Finish:** Floor / Polishing Tile Wall / Lacquer Paint Ceiling / Lacquer Paint

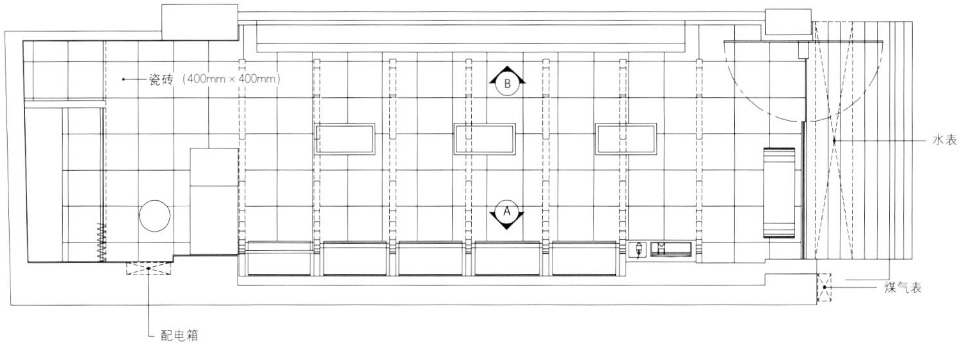

平面图

■ 厅

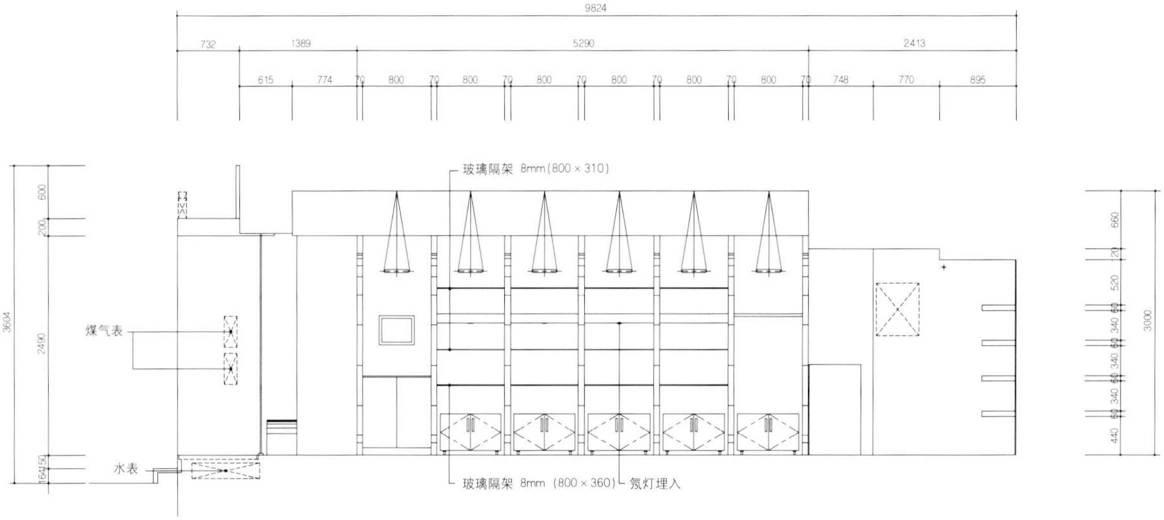

立面图 A（单位：mm）

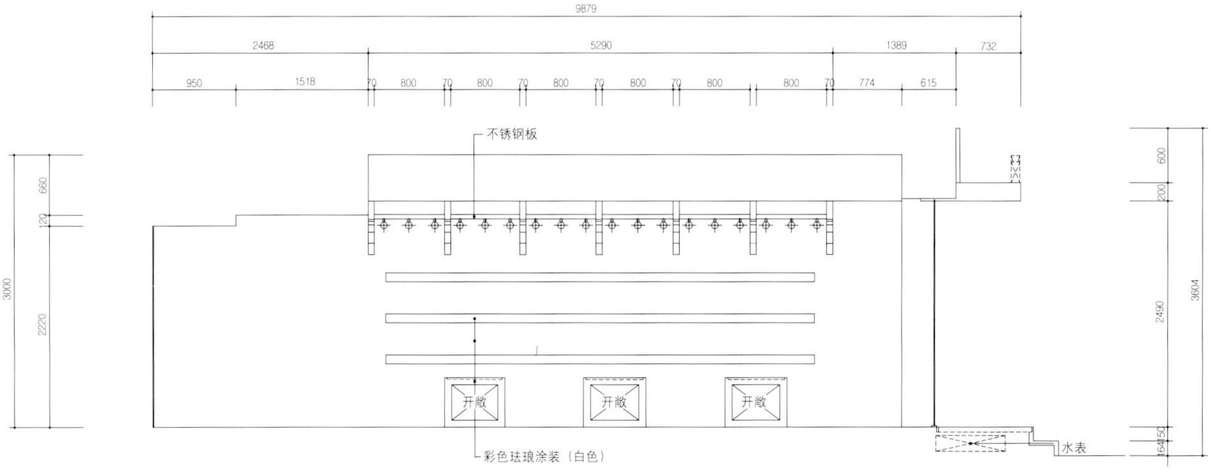

立面图 B

为现代生活方式创造的空间……

这是一处专门为年轻的消费者设计的装饰品分店，位于首尔市西大门。它的设计理念是创造更加幸福的生活。我们都有过走过很长的隧道以后突然有一束光线射入，感到刺眼兴奋的经历，我们也是这样对空间和颜色进行的划分。四边形作为母体以固定的间距对空间进行分隔，用色彩进行强调加工，用所感受到的现代美学进行解析设计。为此我们遵循了分隔的基本原则，综合考虑各种可能性，然后在这上面就像画水彩画一样加上一层透明的颜色。

为了达到实现更幸福生活所进行的设计……

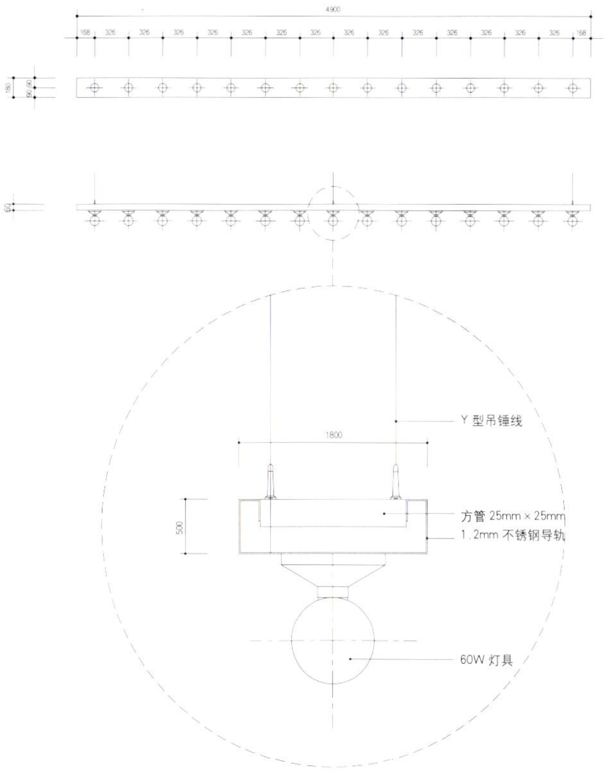

顶棚照明—细部图

■ 展示货架

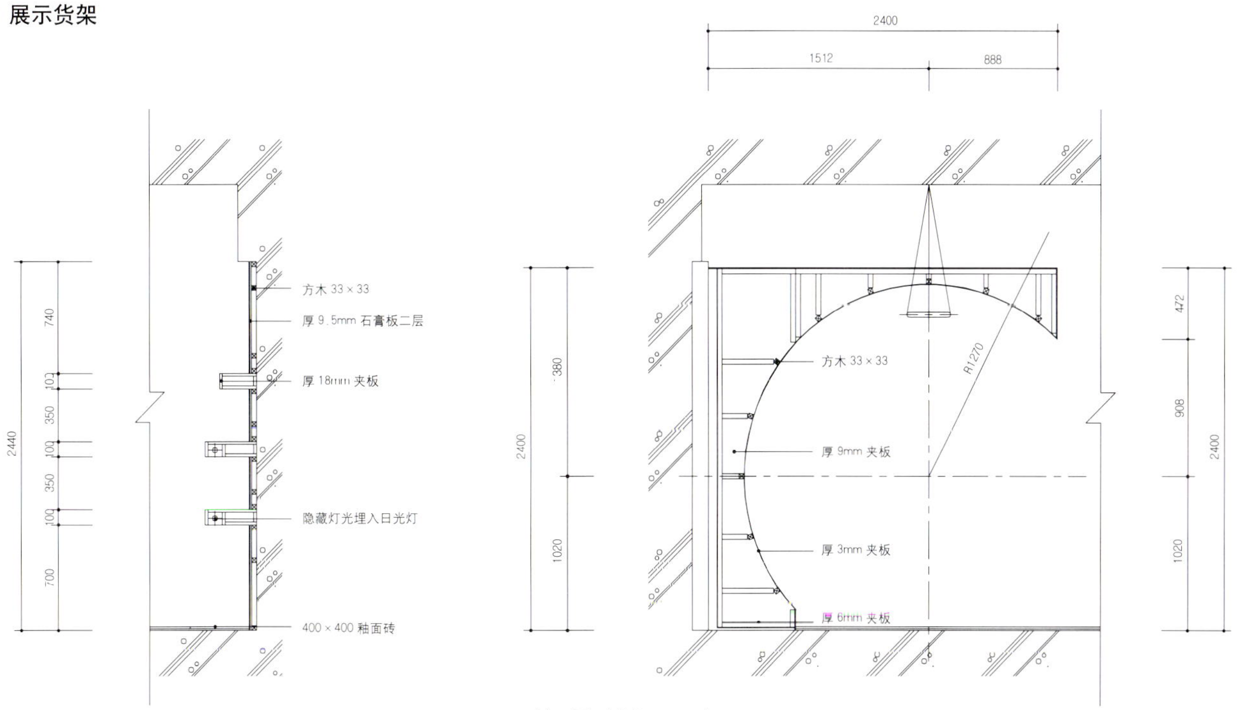

剖面图（单位：mm）

BODA for your life style

With a concept of a more(Boda in Korean) joyful life, the space was designed in expectation of being a beam of light at the end of the tunnel by space/color division. A rectangular space was divided at regular intervals and an element of color was added to produce the beauty of form, which was reinterpreted. Thus we approached with an element of division, taking probability into consideration, and started to draw on it as if doing a watercolor painting. To realize a design for a more joyful life...

Ziller Zone, Daehangno

极乐歌区大学路分店

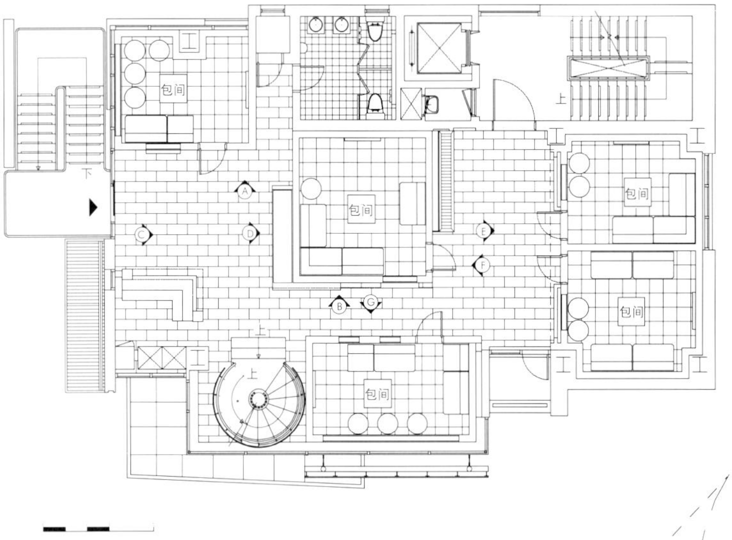

二层平面图

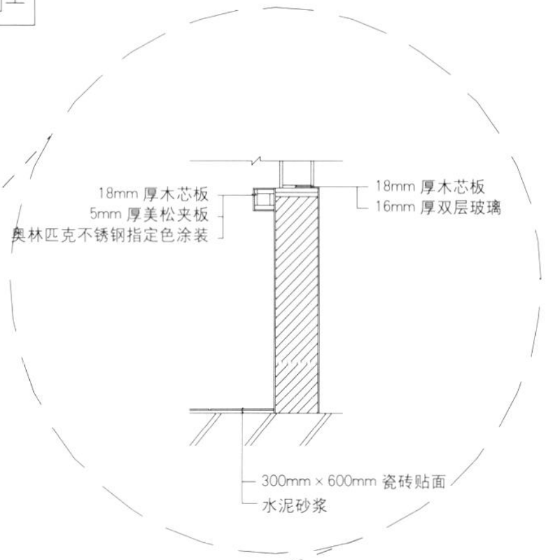

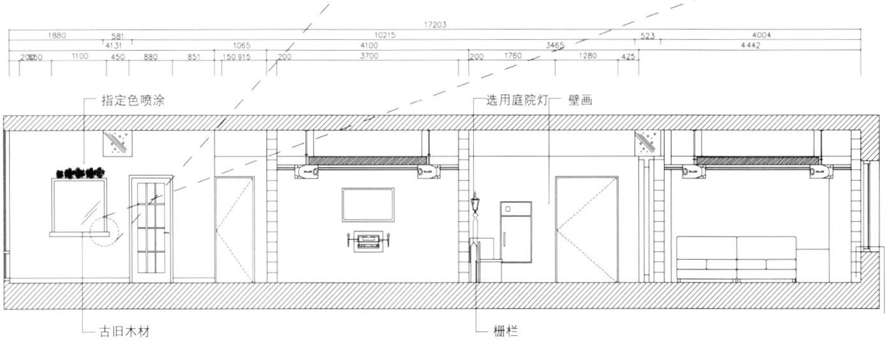

走廊一立面图 A

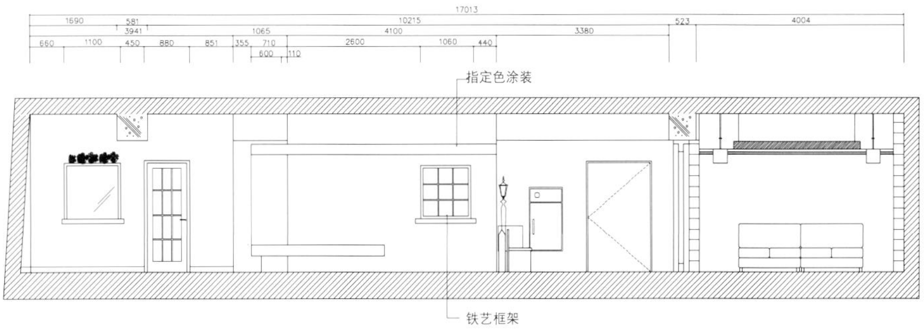

走廊一立面图 B

位于东崇洞的“极乐歌区”是考虑了项目特殊的地理位置而设计的，该处是首尔市文化特区，按照项目建筑规划应该是复合型媒体空间（地下一层为话剧表演、地上一层是设计材料商店、地上二层和三层是练歌房）。

与以前的边饮酒边唱歌的方式不同，这里没有采用古典的欧洲风格，而是以明亮轻快的装修风格为主。

室内统一用米黄色作底色以强调整洁，局部红色作点缀，既有现代感又具有欧洲街道的主题效果。

“明亮轻快的大街……”是这里所要表达的主要风格。为了保持与地下层和地面一层的统一，二层和三层也延续了这种风格。通过使用反射灯和红色系列的亮色，强调了室内空间感，加深了人们对空间的感受。与以往的练歌房给人的感受完全不同，包厢内部的主色调是橘黄色的，沙发则用绿色进行点缀。玻璃材质的旋转楼梯全部用编织布包了起来，通过采用灵活的灯光以及编织布的反射轮廓大大地降低了空间的封闭感，给人以典雅又不失神秘的感觉。路上行人从室外看到灯光打到编织布上的剪影，感受到独特别致的造型美。内部人员可以感受到典雅神秘的氛围。

企划设计：朋友设计公司 权日全 （02）777−9914 | **设计**：朋友设计公司 金秀莲，邱孝正 | **施工**：朋友设计公司 | **地址**：首尔市钟路区东崇洞 1−74TJM 大厦 | 面积：293.3m² | **装修**：地面／釉面砖 墙面／灰墙，彩色涂装，碎砖块 顶棚／水性喷涂，彩色喷涂

Planning Design: Friend's Design · Gwon, Il gwon | **Design:** Friend's Design · Kim, Su ryeon / Gu, Hyo jeong | **Construction:** Friend's Design | **Built Area:** 293.3m² | **Finish:** Floor / Polishing Tile Wall / Lime Plastered Wall, Lacquer Paint, Used Brick Ceiling / Water Spray Paint, Lacquer Paint

■ 二层走廊

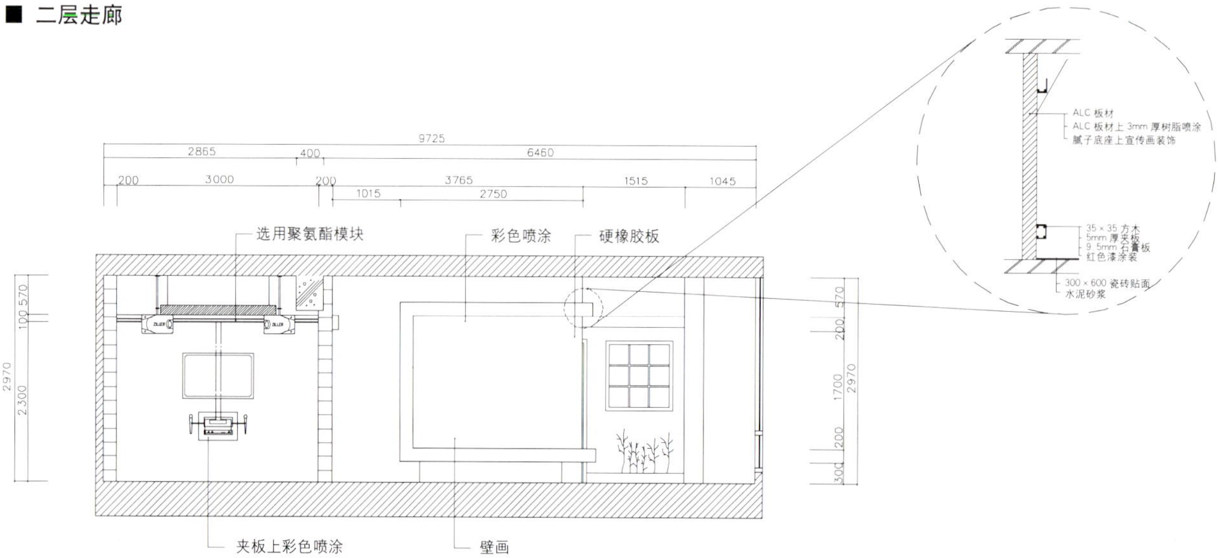

立面图 C（单位：mm）

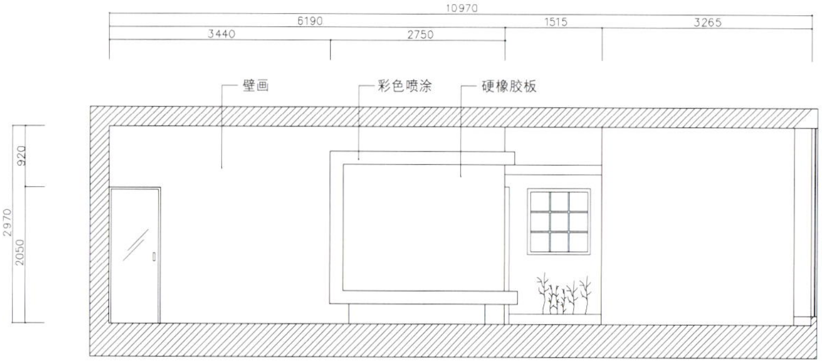

立面图 D（单位：mm）

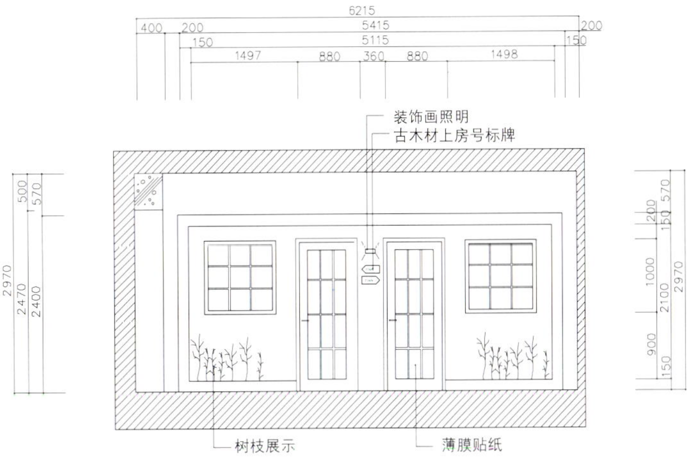

立面图E（单位：mm）

Ziller zone in Dongsung-dong is a complex space reflecting the character of the cultural zone, composed of a live concert hall on the underground, a designer-made accessories shop on the 1st floor, and a singing room on the 2nd and 3rd floor.

Emerging from the existing image of a drinking place, the design put on a classical European style, brightening and cheering the mood. Beautiful colors intensify the neatness and strong red tones accentuate the space, setting a modern and urban European street as a motif.

"A street of a bright, cheerful feeling..." The 2nd and 3rd floor gained a unity from sharing the same concept, and its indirect lighting and strong red tone accentuate the form and spatiality. Inside the room, sofas in main color orange and point color green give a distinction that cannot be found in usual singing rooms. Fabric encases the spiral glass staircase. The space was designed with a cozy and mysterious feeling, its closed feeling minimized by a see-through fabric. To people outside, we attempted to give formative beauty and uniqueness through the moving silhouettes of occupants, and to the people inside the coziness and mystery.

■ 二层

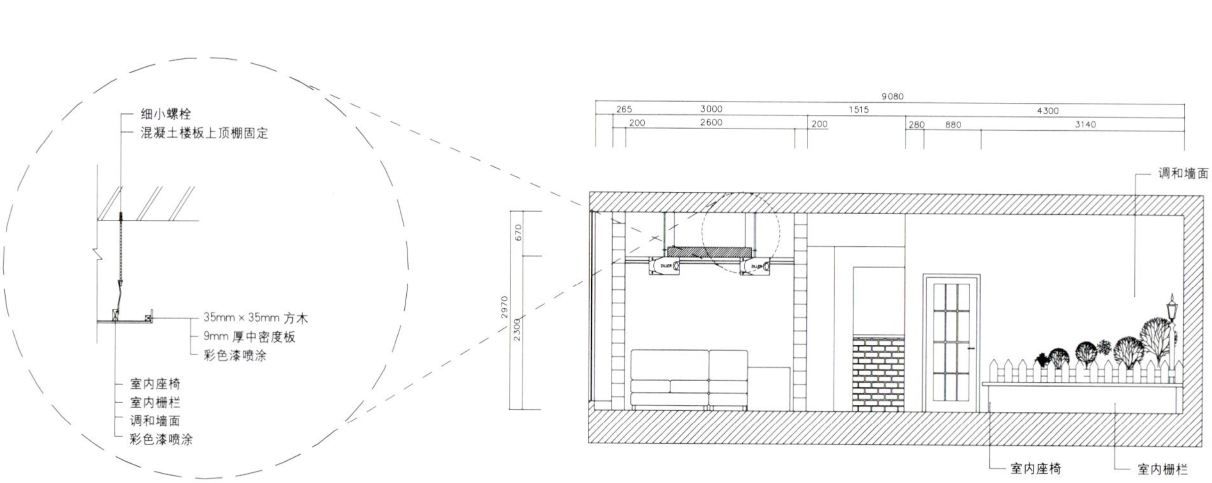
细小螺栓
混凝土楼板上顶棚固定
35mm × 35mm 方木
9mm 厚中密度板
彩色漆喷涂
室内座椅
室内栅栏
调和墙面
彩色漆喷涂
9080
265
3000
1515
4300
200
2600
200
280
880
3140
调和墙面
670
2970
2300
室内座椅
室内栅栏

立面图 F（单位：mm）

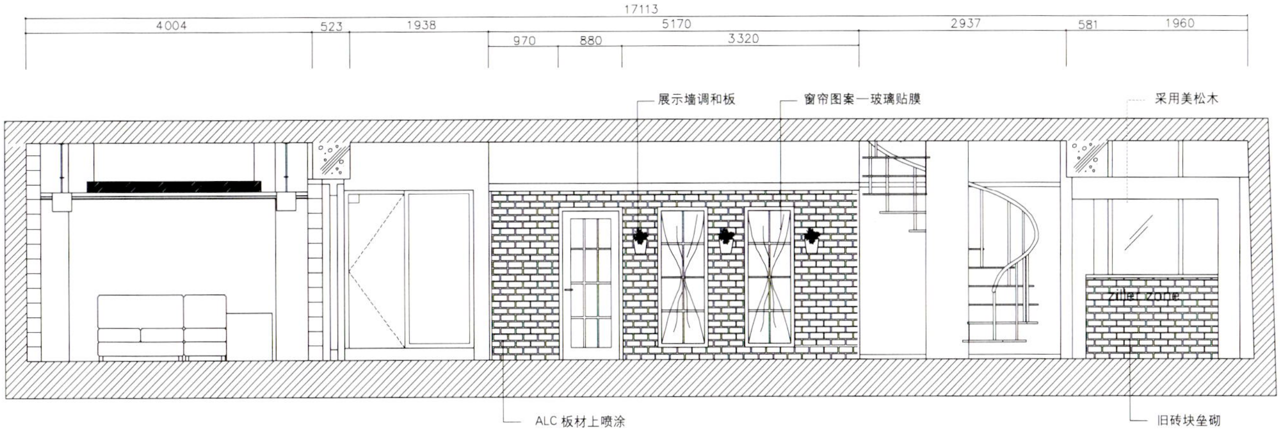

立面图G（单位：mm）

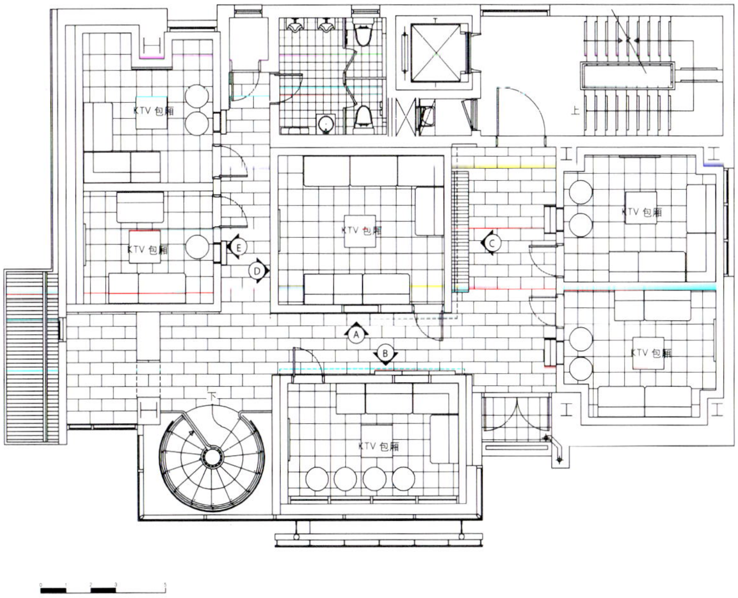

三层平面图

■ 三层走廊

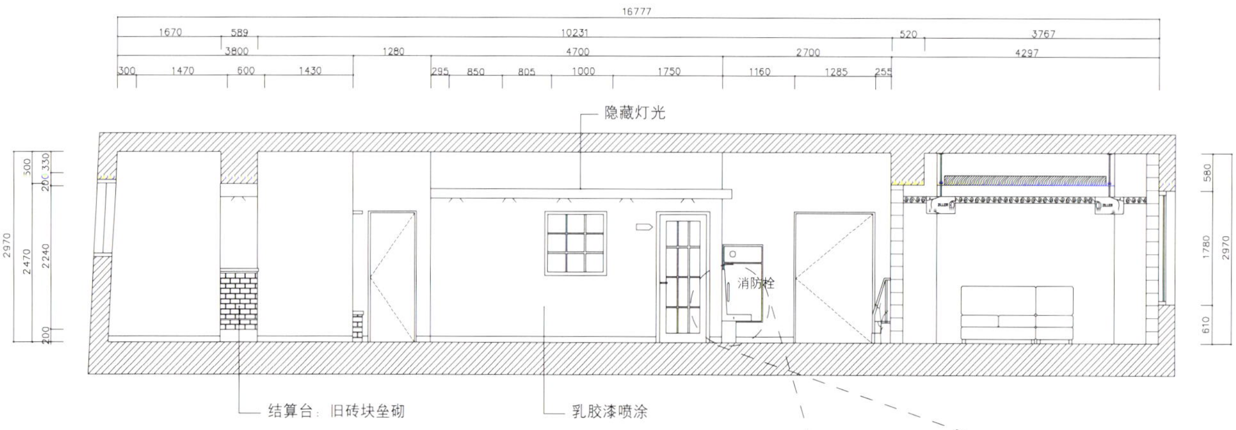

立面图A（单位：mm）

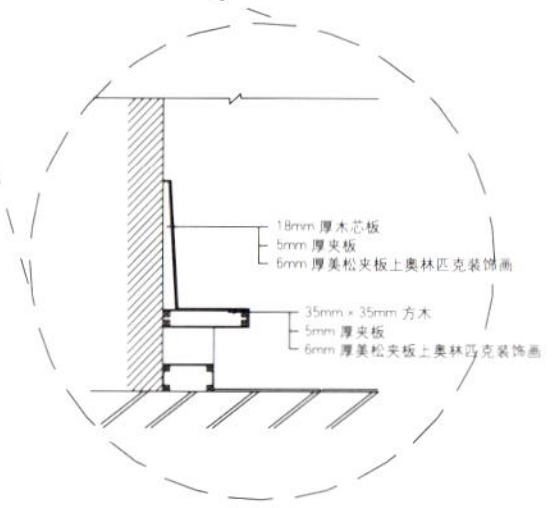

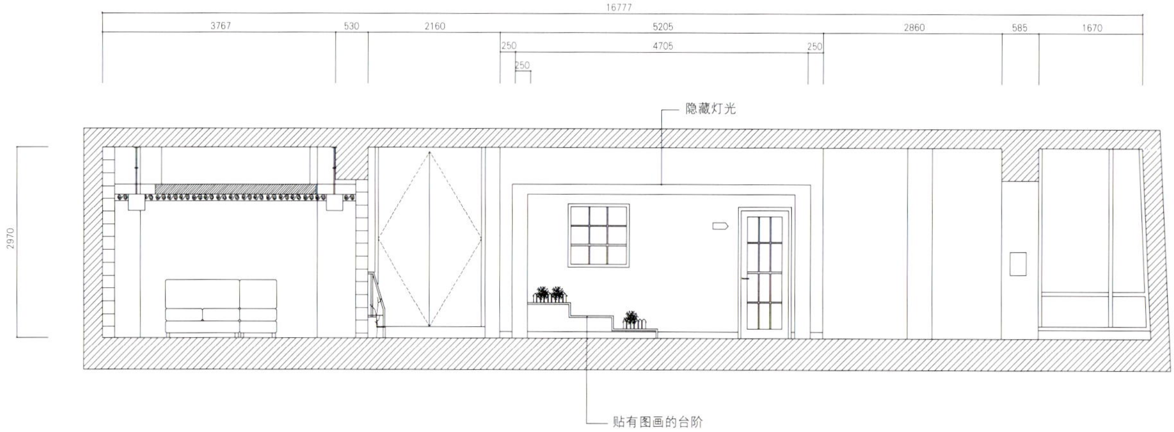

立面图B（单位：mm）

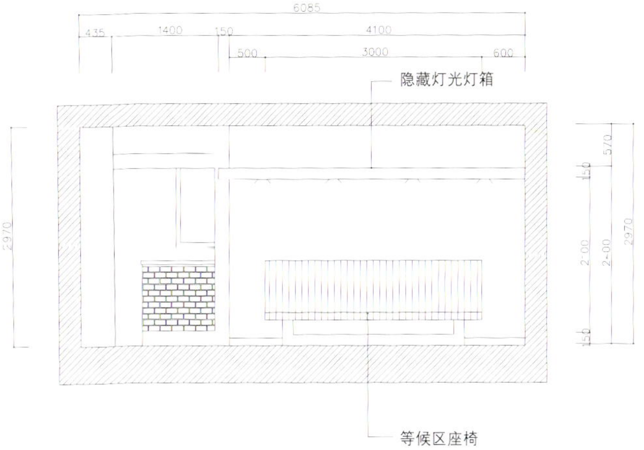

立面图 C（单位：mm）

立面图 D（单位：mm）

立面图 E（单位：mm）

著作权合同登记图字：01－2005－6184号

图书在版编目(CIP)数据

商业建筑／（韩）建筑世界株式会社编；田华等译．—北京：中国建筑工业出版社，2008
最新室内细部设计实例集VI
ISBN 978－7－112－08173－8

Ⅰ.商…　Ⅱ.①建…　②田…　Ⅲ.商业－服务建筑－室内装饰－建筑设计－韩国－图集　Ⅳ.TU247－64

中国版本图书馆CIP数据核字（2008）第006070号

Copyright © 2004 by Archiworld Co., Ltd
Chinese Translation Copyright © 2008 China Architecture & Building Press
All rights reserved.

本书由韩国建筑世界株式会社授权翻译、出版
Interior Detail/Commerce/Archiworld Co., Ltd.

责任编辑：白玉美　戚琳琳　孙　炼
责任设计：郑秋菊
责任校对：王　爽　关　健

最新室内细部设计实例集 Ⅵ
商业建筑
[韩] 建筑世界株式会社　编
田　华　白昭薰　张小卿　郝赤彪　译
*
中国建筑工业出版社出版、发行（北京西郊百万庄）
各地新华书店、建筑书店经销
北京嘉泰利德公司制版
北京方嘉彩色印刷有限责任公司印刷
*
开本：850×1168毫米　1/16　印张：11　字数：310千字
2008年5月第一版　2008年5月第一次印刷
定价：98.00元
ISBN 978－7－112－08173－8
(14127)
版权所有　翻印必究
如有印装质量问题，可寄本社退换
（邮政编码 100037）